The Water Kingdom

also by philip ball

Designing the Molecular World:
Chemistry at the Frontier

Made to Measure:
New Materials for the 21st Century

H_2O:
A Biography of Water

The Self-Made Tapestry:
Pattern Formation in Nature

Bright Earth:
The Invention of Colour

Stories of the Invisible:
A Guided Tour of Molecules

The Ingredients:
A Guided Tour of the Elements

Critical Mass:
How One Thing Leads to Another

Elegant Solutions:
Ten Beautiful Experiments in Chemistry

The Devil's Doctor:
Paracelsus and World of Renaissance Magic and Science

Nature's Patterns:
A Tapestry in Three Parts

Universe of Stone:
Chartres Cathedral and the Triumph of the Medieval Mind

The Sun and Moon Corrupted

The Music Instinct:
How Music Works and Why We Can't Do Without It

Unnatural:
The Heretical Idea of Making People

Curiosity:
How Science Became Interested in Everything

Serving the Reich:
The Struggle for the Soul of Physics under Hitler

Invisible:
The History of the Unseen from Plato to Particle Physics

The Water Kingdom

A SECRET HISTORY OF CHINA

Philip Ball

THE UNIVERSITY OF CHICAGO PRESS
CHICAGO

The University of Chicago Press, Chicago 60637

Published 2017

Printed in the United States of America

26 25 24 23 22 21 20 19 18 17   1 2 3 4 5

ISBN-13: 978-0-226-36920-4 (cloth)
ISBN-13: 978-0-226-47092-4 (e-book)
DOI: 10.7208/chicago/9780226470924.001.0001

Originally published by The Bodley Head, 2016

Library of Congress Cataloging-in-Publication Data

Names: Ball, Philip, 1962– author.
Title: The water kingdom : a secret history of China / Philip Ball.
Description: Chicago : The University of Chicago Press, 2017. |
Includes bibliographical references and index.
Identifiers: LCCN 2016042794 | ISBN 9780226369204 (cloth : alk. paper) |
ISBN 9780226470924 (e-book)
Subjects: LCSH: China—Civilization—Environmental aspects. |
China—Geography. | Water and civilization.
Classification: LCC DS727 .B355 2017 | DDC 951—dc23 LC record
available at https://lccn.loc.gov/2016042794

♾ This paper meets the requirements of
ANSI/NISO Z39.48-1992 (Permanence of Paper).

For Mei Lan
Beautiful Mountain Mist

Contents

Introduction: Rain on the Summer Palace

It was in Beijing during my first visit to China in 1992 that I was introduced to Dr Zhang. To my shame I never kept a record of his full name, and I can't hope to discover it now: you might as well seek a Dr Smith in London or a Dr Schmidt in Munich. I do know that Dr Zhang worked at the Beijing Institute of Electron Microscopy, and when we met I was relieved to find that his English was very good, because at that time my Chinese would not have taken me beyond a salutary 'Ni hao!'

Doing scientific research in China in those days was fraught with mundane difficulties. As Dr Zhang showed me around his laboratory there was a power cut, and the graduate students filed out of the building with a mixture of exasperation and resignation. 'Won't it damage the microscopes?' I asked. 'Oh yes,' Dr Zhang replied nonchalantly.

When he wondered what else I had seen in Beijing, I confessed that my sightseeing had hitherto been almost non-existent. My days had been filled with visits, obligations and dauntingly copious meals with my academic hosts. Then we must go to the Summer Palace, Dr Zhang insisted. We hopped aboard an overcrowded bus that made its slow progress past fleets of bicycles – the only cars on the Beijing roads back then were state-owned taxis – to the outskirts of the city.

It was raining a little, which had the happy consequence that the Summer Palace was relatively quiet. I welcomed the rain for other reasons too. Dripping from the dragon-decorated porticos, it was emblematic of my romantic visions of China, which had otherwise taken something of a battering since my arrival. Little boys called out

'Hello!' and tried to sell me cobs of roasted sweetcorn. 'That's their one word of English,' said Dr Zhang. 'That, and "one yuan". Every old lady knows those words.'

I was profoundly glad that I had a guide to take me around the Summer Palace, since otherwise I would have missed so much. It owes its present form, he told me, to the formidable Empress Dowager Cixi, who had it rebuilt in the late nineteenth century during the reign of her nephew the Guangxu Emperor – although she, as regent, should have spent the money instead on a navy or on healing the ailing Qing economy. The palace is a maze of symbols, obvious to the Chinese but invisible to Western eyes. Towers and pagodas are octagonal, Dr Zhang explained, because eight is a special number in China, indicating good fortune and 'old money'. Someone who has a good manner with all kinds of people is said to possess 'eight ways'. (I suspected Dr Zhang was one of these.) But the base is square, symbol-izing the earth, and the top is round or has round designs painted on it, representing the sky.

We strolled along a covered wooden walkway, a pagoda-topped hill on one side and a lake on the other. It looked lovely, but there was more to the view than that. The hill indicates (if my diaries can be trusted) prosperity, the water means long life. A view of the two at once – hill and lake – is harmonious. I did not know it then, but the Chinese word for 'landscape' is simply the combination of the two: *shanshui*, mountains (*shan*) and water (*shui*).

We came to an impressive building that was used as a library of sorts, an attempt by the empress to show that the imperial family was cultured. On the front of the building was an inscription.

'This is . . .' said Dr Zhang, about to translate for me. Then he hesitated. 'Er, I don't know what you would say in English.'

I was surprised, since his English seemed so good. He eventually came up with something like 'Between mountains, water and sky one feels a concentration of harmony.' But he was evidently not very happy with this version.

'It is really like a feeling,' he said. 'It is like the way these trees are placed here – it conjures up an atmosphere or emotion.' Yes it did, in the soft rain, despite the groups of Chinese visitors posing for photo-graphs behind us, flashing the ubiquitous splayed-finger 'V' imported from the West and apparently emptied of meaning.

'Chinese paintings are the same,' he went on. 'They are about feeling, not about realism.' I was suddenly aware that my guide and host, who spent his days looking into microscopes and hoping the power didn't cut out, was a man with a finely developed aesthetic sense – one that, moreover, my own language and culture were ill-equipped to give to me.

Anyone from outside China who begins to learn the language and to wrestle with this rich and subtle culture comes sooner or later to realize that decoding the characters and tones is only a small part of the battle. There are of course limits to translation between any two cultures, but words in Chinese are not just semantic signifiers. They are distillates of Chinese thought, saturated with association and ambiguity, ready to unfold layers of meaning that differ according to context. That is why it is better in any case to call them characters, not words: their form and content are inextricably entwined. The little four-character sayings that Chinese people love to quote to the bafflement even of the foreign student who can decrypt the individual meanings are a kind of philosophical distillate: ideas, stories, legends, concentrated beyond reach of literal translation. I know now, as I didn't in 1992, why it was so hard for Dr Zhang to find the right words for what we saw at the Summer Palace. Recalling that damp afternoon reminds me too that an attempt by a *waiguoren*, a foreigner, to explain what a concept like water means within Chinese culture is doomed from the outset to all manner of oversights, failures of nuance and, saddest of all, failures of imagination.

What makes me attempt it nonetheless is precisely the fact that I am from the far-off lands long dismissed by the Chinese as the realm of barbarians. I suppose I have in my defence the hoary old quote attributed to the anthropologist Ralph Linton that 'the last thing a fish would ever notice would be water', except that this is not quite the right way to describe the situation of the Chinese people. It is precisely *because* they notice water in so many ways, with such sophistication and connotation, that it is perhaps harder for them to recognize how deeply and widely water has influenced and defined China. It's just too obvious.

Water, as Dr Zhang revealed to me that rainy summer day, is one of the most powerful vehicles for Chinese thought. At the same time, and for the same reasons, it has been one of the key determinants of Chinese civilization. It has governed the fates of emperors, shaped the contours of Chinese philosophy, and left its mark, quite literally, throughout the Chinese language. It is with water that heaven communicates its judgements to earth. Water pronounces on the right to govern. For these reasons, there is no better medium for conveying to the barbarians beyond the Wall what is special, astonishing, beautiful, and at times terrifying and maddening, about the land its inhabitants call Zhongguo, the 'Middle Kingdom' – which is in the end the Water Kingdom, too.

I hope that Chinese readers might also appreciate and enjoy this *waiguoren's* glimpse at their culture. I even dare hope that Dr Zhang – if he is still peering into microscopes, if his eyes have not failed him – might recall that day too. I am grateful for it.

Opening a window on the Middle Kingdom

Here is the original *waiguoren* marvelling at China's extraordinary accomplishments – Marco Polo, in his contested *A Description of the World* (c.1300):

> I repeat that everything appertaining to this city is on so vast a scale, and the Great Khan's yearly revenues therefrom are so immense, that it is not easy even to put it in writing, and it seems past belief to one who merely hears it told.

As Marco Polo testified, the sheer magnitude of everything about China confounds the imagination of the outsider. If that was true in the thirteenth century, when the Venetian explorer might or might not (opinion remains divided) have witnessed at first hand the court of Khubilai Khan, the Mongolian first emperor of the Yuan dynasty, it is even more so today. One now has to try to comprehend not just the vastness of the country, the diversity of its population, geography and languages, and the antiquity of its culture, but also the scope and ambition of its global economic and political status and the scale of the consequent internal and external upheavals that generates. Countless books today offer to reveal 'how China works' or 'what China's future holds'. But anyone familiar with the country will recognize that not even the Chinese truly know these things, and that the answers you get depend – as they always do in China – on whom you ask.

The purpose of *The Water Kingdom* is different. In exploring one of the most constant, significant and illuminating themes among the turbulent and often confusing currents of Chinese history and culture – that is to say, the role of water – it seeks to show how the nation's philosophy, history, politics, administration, economics and art are intimately connected to a degree unmatched anywhere else in the world. For this reason, it is not simply the case that all these facets of Chinese culture become easier to understand when the role of water is recognized. Rather, one must conclude that many of them are likely to remain strange, opaque or alien *unless* their connection with water is understood.

In this way, the grand journey that *The Water Kingdom* takes through China past and present opens a window through which one can begin to grasp the potentially overwhelming complexity and teeming energy of the country and its people. Water is a key that unlocks an extraordinary quantity of Chinese history and thought. One could tell the political and economic history of the nation largely through the medium of water, and indeed the historian Ch'ao-ting Chi (Ji Chaoding) has done so in his influential *Key Economic Areas in Chinese History* (1936). Chi's subtitle said it all: 'as revealed in the development of public works for water-control'. For Chi, shifts in the major economic areas of China throughout its imperial history depended on the state exploitation of waterways, both natural and artificial, for agriculture and military transportation.

Of course, there are limits to and blind spots in that approach. One could argue that political power, social stability and a cultural sense of place have been rooted more in notions of land than water: after all, water is ultimately a means to the end of growing and transporting the crops on which life in China has long depended. But the importance of water for agriculture, transport and social stability has made it a central element of political power, and has shaped the way that the country has been governed. Many of the challenges that water poses are much the same today as they were for the ancient Han emperors, so that, even with the opportunities that new technology provides, the Communist Party has not always been able to distance itself from China's imperial past. Indeed, the government is now finding that water is a useful vehicle for shrewd mobilization of that heritage. Water offers a language, universally understood within China, for asserting and articulating political legitimacy, expressed since ancient times in the form of engineering works on a scale that no other nation on the planet could have contemplated – for better and worse.

All this tells just one side of China's relationship with water, namely the hydraulic and hydrological foundations of state power. Yet while water serves as a central symbol to emperors, governors and party chairmen, it also has profound meanings for the ordinary people. It is all too easy to give the impression that China's population has been so many powerless individuals moved hither and thither by their leaders. It has sometimes felt that way for these people too. But the ubiquitous and ambivalent relationship that the Chinese people have had with water has made it a powerful and versatile metaphor for philosophical thought and artistic expression, and its political connotations can be subverted and manipulated in subtle ways for the purposes of protest and dissent. These meanings of water are more than metaphorical. Because the lives of everyday folk have always depended on water, the rivers and canals mediate their relationship to the state. Water – too much of it, or too little – has incited the people to rise up and overthrow their governments and emperors. Burgeoning economic growth now places unprecedented pressure on the integrity and sometimes even the very existence of China's waterways and lakes. Not only can China's leaders ill afford to ignore this potential brake on economic growth, but the environmental problems are leading to more political pluralism in a nominally one-party state.

To tell this tale, *The Water Kingdom* must offer a different kind of journey from that of Marco Polo, which still sets the tenor of many Western accounts in presenting China as a place of baffling size and strangeness. Our guides are instead often Chinese travellers and explorers, poets and painters, philosophers, bureaucrats and activists, who have themselves struggled to come to terms with what it means to live within a world so shaped and permeated by water. You do not need to take my word for it that water is so central to Chinese culture: you have theirs.

Finding your way

It starts with the Qin in the third century BC, and ends with the Qing at the close of the nineteenth century, and there already you have the problem for the non-Chinese non-expert, who will struggle even to hear the difference between these two dynastic names that bookend China's documented imperial history. On the way, one has to negotiate the states of Qi and Qu, not to mention more than one Jin, and to appreciate that the Han dynasty is not the same as the state of Han squashed between formidable Qin, Wei and Chu during the Warring States period, before the Qin unified the land. Names were recycled, sometimes to imply historical links that conferred status and legitimacy. And while the distinctions between lands blur into monosyllabic confusion, emperors become hard to keep in mind for the opposite reason that their names proliferate perplexingly (see below).

For the average Westerner, 'Ming' represents a porcelain vase of a historical provenance that is vague but distant enough so as to make the vase worth a few bucks. Meanwhile, Tang, Song and Sui are at best interchangeable designations of antiquity that conjure up little more than Tony Cheung and Jet Li crossing swords. I apologize at once both to those who have bothered to get a little (perhaps a lot) more acquainted with China's past, and to Chinese people for whom this Western ignorance of a vast and sophisticated heritage is rightly seen as nothing less than shameless insularity. In 1992 I was in the same place, standing in Tiananmen Square having managed to peruse (and not really absorb) little more than the primer at the beginning of the *Lonely Planet* guide.

I hope that *The Water Kingdom* will not, for novices to Chinese history, seem too often like another of these litanies of names that leaves the mind struggling to find purchase. I have at the very least a duty to try to give some orientation from the outset. A paragraph or two summarizing China's dynastic history is sure to be tragically crude, but might at least offer a reference point to which you can return if you have forgotten where Sui sits in relation to Song. It really does not take so long before these dynastic labels cease to be empty syllables and begin to acquire the kind of period flavour that for the English is conjured up by 'Plantagenet', 'Tudor' and 'Victorian', for the Americans by 'Washington', 'Lincoln' and 'Eisenhower', or for the Italians by 'Medici', 'Garibaldi' and 'Mussolini'.

Dynastic succession has traditionally supplied the schema by which Chinese historians of earlier times have told their country's past. It offers a framework, a kind of aide-memoire, on which to hang a narrative that is often every bit as artificial as the way the Renaissance historians invented the Dark and Middle Ages to exalt their own supposed links to classical antiquity. There was rarely if ever handing on of power with the relative smoothness of the transition from Tudor to Stuart, or from Stuart to Hanover. The edge of empire might shift significantly between dynasties, and in some instances a new dynasty (such as the Jin or the Yuan) was created by conquerors from outside of the Middle Kingdom. Some eras come with a dynastic label attached even though parallel designations confess to a time of disunity elsewhere in China – so that, say, the Eastern Jin dynasty more or less coincides in the fourth and fifth centuries with the Sixteen Kingdoms period. With these warnings in mind, here is the 'official picture':

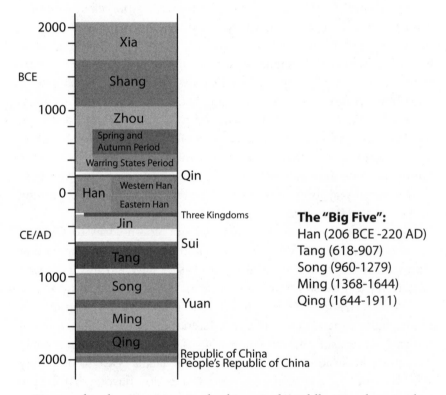

The "Big Five":
Han (206 BCE -220 AD)
Tang (618-907)
Song (960-1279)
Ming (1368-1644)
Qing (1644-1911)

Geographical orientation in the historical Middle Kingdom is also complicated, even more than it is in many other parts of the world, by the shifting boundaries of kingdoms and empires and the sometimes almost casual renaming of cities, which was among other things a way for the founder of a new dynasty to impose his authority. Occasionally, cities that grew and merged were awarded composite names, much in the manner of Budapest: the Yangtze conurbation of Wuhan is the archetypal case. In some instances I will indicate the old names of cities that have different labels today, with the proviso that the old and new are not necessarily coincident. But I make no claim or promise to be systematic about this: what I might sacrifice in clarity that way I hope to excuse as the avoidance of pedantry.

On language

Every Western book on China is obliged to include a prefatory note explaining the chosen scheme for romanization of the language. This one is no exception. These choices always have an arbitrary aspect to them. Advocates of the modern Pinyin system, like me, are liable to dismiss the older Wade–Giles system ('Mao Tse-tung' and so forth) on the basis that it accrues a confusing proliferation of accidentals while dishing out countless little phonetic insults. ('Peking' bears rather little resemblance to the way the current inhabitants of that city pronounce its name.) Others accuse Pinyin (with some justification) of seeming to make selections from the Roman alphabet almost as if to redeem the most neglected letters, while assigning them sounds barely related to those of Western use (so that *q* becomes *ch*, for example). In any event, Pinyin is what I shall mostly use; I hope that I will not upset too many sinologists by the inconsistency of retaining a few older romanizations that have become familiar to Western eyes, so that we have Yangtze instead of Yangzi, and Lao Tzu instead of Laozi.

For those unfamiliar with this system, here is a very rough guide to the most salient matters of pronunciation:

q: like *ch* in 'chat'
x: close to the German *sch*
zh: like *g* in 'gentle'
z: like *dds* in 'adds'
c: like *ts* in 'cats'
uo: not *oo-oh*, but more like *wo* in 'woke'
e, en, eng: like *uh, uhn, uhng*
o in *bo, mo,* etc.: more like *aw* in 'saw' than *oh* in 'so'

ai: like 'aye'
ao: like *ow* in 'cow'
ei: like *ay* in 'way'
ui: like 'way'
iu: like *yo* in 'yo-yo'
i in *zhi*, *si*, *ci*, *ri*, *shi*: like *e* in 'the'
i in *ni*, *yi*, *bi*: like *ea* in 'tea'
ie: like *ere* in 'here'
j: always hard, as in 'jump' (even in *Beijing*!)

I must add a cautionary note on naming conventions. Even setting aside the complications of different romanization systems, the naming of Chinese emperors and officials is far from straightforward. There are often several alternative appellations to choose from. For example, the sage-king Yao, one of the semi-mythical Five Emperors, also has a clan name (Taotang) and a given name (Fangxun), and he may also be called Tang Yao. Emperors had a 'temple name' as well as a family name, and up to and including the Song dynasty they generally combined their temple name with the suffix *di* (simply meaning emperor or ruler), *zu* (generally used by the founder of a dynasty) or *zong*.

Emperors would also employ 'era names', which described the nature of particular phases of their rule; for example, the Han Emperor Wudi's first era name was Jianyuan, meaning 'establishing the first'. Until the Ming dynasty, emperors might change their era name several times during their reign (obliging calendars to be reset to Year One each time), which is why they are identified by their temple name. The Ming and Qing emperors, however, chose an era name that lasted for the duration of their rule, and so it is by this name that they are known. The peasant rebel Zhu Yuanzhang who founded the Ming dynasty in the fourteenth century, for example, became the Hongwu ('vastly martial') emperor, and also took the temple name Ming Taizi. Since era names are not exactly names in the usual sense, one should properly refer to them in this manner: the Hongwu and Qianlong emperors, say. In general I will observe that convention, but occasionally I will flout it for brevity.

It's not just for emperors that naming is complicated. Many Chinese people, even into the twentieth century, followed the widespread East

Asian custom of adopting a 'courtesy name' when they reached adulthood: the poet Du Fu took the courtesy name Zimei, for example, while Mao Zedong was Runzhi, and Zhu Yuanzhang was Guorui. These names were respected: the twelfth-century official Lu You meticulously records the courtesy name of all the various dignitaries whom he encountered on his travels. I must regretfully omit that courtesy.

1 The Great Rivers

Yangtze and Yellow: The Axes of China's Geography

The wide, wide Yangtze, dragons in deep pools;
Wave blossoms, purest white, leap to the sky.

Lu You (1125–1209),
'The Merchant's Joy'

The sun goes down behind the mountains;
The Yellow River flows seaward.
You can enjoy a grander sight
By climbing up one floor.

Wang Zhihuan (688–742),
'At Heron Lodge'

When Confucius described water as 'twisting around ten thousand times but always going eastward', he seemed to imply that the east-bound flow of rivers was tantamount to a law of nature, almost a moral precept. There is no clearer illustration of how a culture's geography may affect its world view. Why would anyone who had never stepped foot outside China have any reason to doubt that this was how the world was made?

In China the symmetry of east and west is broken by tectonic forces. Westwards lie the mountains, the great Tibetan plateau at the roof of the world, pushed upwards where the Indo-Australian plate crashes into and plunges beneath the Eurasian. Eastward lies the ocean:

only Taiwan and Japan block the way to the Pacific's expanse, which might as well be endless. The flow, the pull, the tilt of the world, is from mountains to water, from *shan* to *shui*.

This is the direction of the mighty waterways that have dominated the country's topographic consciousness. 'A great man', wrote the Ming scholar and explorer Xu Xiake, 'should in the morning be at the blue sea, and in the evening at Mount Cangwu' (a sacred peak in southern Hunan province). To the perplexity of Western observers (not least when confronted with Chinese maps), the innate mental compass of the Chinese points not north–south, but east–west. The Chinese people articulate and imagine space differently from Westerners – and no wonder.

All of China's great rivers respect this axis. But two in particular are symbols of the nation and the keys to its fate: the Yangtze and the Yellow River. These great waterways orient China's efforts to comprehend itself, and they explain a great deal about the social, economic and geographical organization of its culture and trade. The rivers are where Confucius and Lao Tzu went to think, where poets like Li Bai and Du Fu went to find words to fit their melancholy, where painters discerned in the many moods of water a language of political commentary, where China's pivotal battles were fought, where rulers from the first Qin Emperor to Mao and his successors demonstrated their authority. They are where life happens, and there is really nothing much to be said about China that does not start with a river.

Search for the source

The great rivers drove some of the earliest stirrings of an impulse to explore and understand the world. The Yü Ji Tu ('Tracks of Yü' Map), carved in stone sometime before the twelfth century, shows how Chinese cartography was far ahead of anything in Christendom or classical Greece. In medieval maps of Europe the rivers are schematic ribbons, serpents' tails encroaching from the coast in rather random wiggles. But the Yü Ji Tu could almost be the work of a Victorian surveyor, depicting the known extent of the kingdom with extraordinary fidelity and measured on a very modern-looking grid. It is dominated by the traceries of river networks, with the Yellow River and the Yangtze given bold prominence. These are the 'tracks' defined by China's first

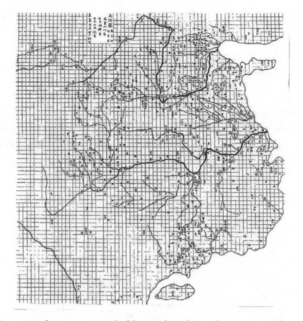

The Yü Ji Tu, carved in stone probably in the eleventh century. The cartographic use of a grid system dates at least back to the Han dynasty in the second century AD, when the polymath Zhang Heng is said to have introduced it.

great water hero, the legendary emperor Yü who conquered the Great Flood (Chapter 2).

China has always been interested in – one might fairly say obsessed with – its rivers. The *Shui jing* (*Classic of the Waterways*) was the canonical text of hydrological geography, traditionally credited to Sang Qin of the Han dynasty, although later scholars have placed it in the third and fourth centuries AD (the Jin dynasty). We don't know quite what it contained, since it has been lost, but a commentary on the work, known as the *Shui jing zhu* by the scholar Li Daoyuan (427–527), ran to forty volumes and listed more than 1,200 rivers.

The impassioned searching for the source of the great rivers throughout Chinese history seems almost to betray a hope that it will reveal the occult wellspring of China itself, the fount of the country's spirit (*qi*). The source of the Yellow River was debated at least since the Tang dynasty of the seventh to the tenth century AD, and the Yuan emperor Khubilai Khan dispatched an expedition in 1280 that was supposed to clarify the matter. Yet the point was still being argued

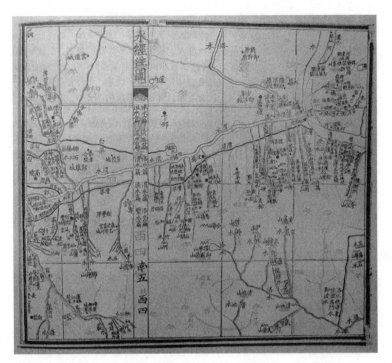

The Wei River as depicted in the fifth-century work *Shui jing zhu* (*Commentary on the Classic of the Waterways*).

seven centuries later, when the China Exploration and Research Society declared that the Yellow River springs from the icy, crystal-clear waters of lakes Gyaring and Ngoring in the Bayan Har Mountains of remote Qinghai.

The source of the Yangtze is disputed even now. An expedition in the 1970s identified it as the Tuotuo, the 'tearful' river in Qinghai, but several years later it was assigned to the Damqu instead. There's ultimately something arbitrary in conferring primacy on one of a river's several headwater sources, but for the Yangtze the symbolic significance of this choice is too strongly felt for the protagonists to brook any compromise. The classical answer, given in the *Yu gong* manuscript from the Warring States period of the fifth to the third century BC, was that the Yangtze begins as the Min River in Sichuan. But during the Ming era, iconoclastic Xu Xiake (1586–1641) argued otherwise. He found that the Jinsha River, which joins the Min in Sichuan, goes back much further than the Min: a full 2,000 kilometres,

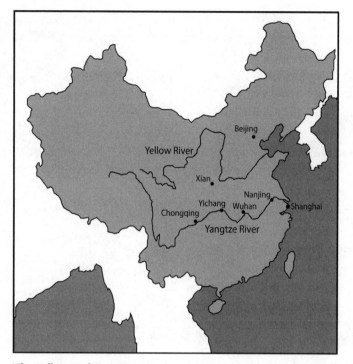

The Yellow and Yangtze rivers.

deep into the wilds of the Qinghai plateau. The Jinsha ('Golden Sand', referring to the alluvial gold that may be found in the river's sediment) itself stems from the Dangtian, whose tributaries in Qinghai vie as the ultimate source of the Yangtze, flowing from the glacier lakes of that high and inhospitable land.

No one better personifies the Chinese devotion to its great rivers than Xu Xiake, who wandered for thirty years into remote places, suffering robberies, sickness, hunger and all manner of hazards. 'He would travel', one contemporary account relates,

> with a servant, or sometimes with a monk and just a staff and cloth bundle, not worrying about carrying a travelling bag or supplies of food. He could endure hunger for several days, eating his fill when he found some food. He could keep walking for several hundred *li*,*

---

* The Chinese *li*, like the European foot, was a measure that varied between eras, but was approximately equal to between a third and a half of a kilometre.

ascending sheer cliffs, braving bamboo thickets, scrambling up and down, hanging over precipices on a rope, as nimble as an ape and as sturdy as an ox. He used towering crags for his bed, streams and gullies for refreshment, and found companionship among fairies, trolls, apes and baboons, with the result that he became unable to think logically and could not speak. However, as soon as we discussed mountain paths, investigated water sources or sought out superior geographical terrain, his mind suddenly became clear again.

From *shui* to *shan*: what more nourishment could the mind need? And to get there, Xu believed, one should not march like a soldier but wander like a poet.

In the person of Xu Xiake, Confucian rectitude meets Daoist instinctiveness and reverie. He was born in the city of Jiangyin, north-west of Shanghai on the Yangtze delta. For much of his travels Xu was attended by a long-suffering servant named Gu Xing. The pair often had to rely on the benevolence of local monasteries for food and shelter, where Xu might offer payment in kind by writing down the history of the institution. On one occasion they were attacked and robbed by bandits on the banks of the Xiang River in Hunan, left destitute but lucky to be alive. Perhaps we can forgive Gu for finally robbing and deserting his master.

Xu journeyed into snowy Sichuan and harsh, perilous Tibet, where rivers could freeze so fast that wandering cattle could get trapped and perish in the ice. He went deep into the steamy Yunnan jungle, then still a region alien, foreign and wild, to determine that the Mekong (called the Lancang in China), Salween (Nu) and Red (Lishe) rivers were separate entities along their entire courses. But although he diligently recorded the local geology and mineralogy, there is little that is systematic in his itinerary: he was wandering more or less without plan or destination.

Still he deserves to be called a geographer. His methods of surveying were crude, but they rejected the local superstitions that until then supplied the usual rationale for natural phenomena. His notes, according to the great scholar of Chinese science and technology Joseph Needham, 'read more like those of a twentieth-century field surveyor than of a seventeenth-century scholar'. And like his contemporaries in Europe, he was prepared to risk censure by preferring the testimony of experience over that of classical authorities. There had been whispers ever since the Han era that the true headwaters of the Yangtze were not, as the classics insisted, the Min, but instead the Jinsha flowing from the Kunlun Mountains of Qinghai. Xu, however, was the first to dare make the claim openly. For this he was denounced as despicable.

Ancient scholarly study of China's rivers and waters reveals how far ahead of the West Chinese theory and practice were, not only in cartography but in an understanding of natural phenomena. While the *Shan hai jing* (*Classic of Mountains and Seas*, probably written in the Warring States period) was content to ascribe the tides to the comings and goings of a massive leviathan-like creature in the oceans, the Han scholar Wang Chong argued in the first century AD that tides are related to the moon. 'The rise of the wave follows the waxing and waning moon,' he wrote, 'smaller and larger, fuller or lesser, never the same.' Wang Chong championed a rationalistic explanation of the world over the rather superstitious Daoism and formulaic Confucianism of his time, and his meteorological and astronomical observations were particularly astute. He described the essence of the hydrological cycle (even if his belief in the link between the moon and water extended to a lunar influence on rainfall): 'Clouds and rain are really the same thing. Water evaporating upwards becomes clouds, which condense into rain, or still further into dew.' Wang Chong perceived the same correspondences between the movements and forms of river water and of blood circulation that were noted by Leonardo da Vinci a millennium and a half later. He wrote:

> Now the rivers in the earth are like the pulsating blood vessels of a man. As the blood flows through them they throb or are still in accordance with their own times and measures. So it is with the rivers. Their rise and fall, their going and coming are like human respiration, like breath coming in and out.

The value of such beliefs, as many historians of science have noted, is not so much a matter of whether or not they are true, as of their capacity to stimulate further observation and to explain the world in naturalistic terms. The importance of the waterways created an imperative for such speculations, just as it drove the development of technologies and systems for making careful measurements and records, for example so that water levels could be determined during dredging operations. Cartography was so far advanced in China from the Han to the Ming eras partly because water management was accorded such priority.

China's Sorrow

What a strange journey the Yellow River, China's 'mother river' (*muqin he*), makes from mountain lake to Yellow Sea. Pouring down from the western highlands, around the city of Lanzhou in Gansu province it departs from its eastwards flow and travels north towards Inner Mongolia, then executes another bend to turn south along the border of Sha'anxi and Shanxi provinces. Finally, sluggish with silt and descending the shallowest of gradients, it turns abruptly east when joined by the Wei River near the border of Sha'anxi and Henan. It cuts north-east across the North China Plain, through Henan and Shandong, before emptying at the coast. The 4,632-kilometre journey makes the Yellow River the fourth longest in the world. The flow is not so massive compared with the Amazon or the Mississippi, but it varies hugely between the dry and wet (June–September) seasons. That is partly what makes the Yellow River so hard to manage – but the key problem is the silt.

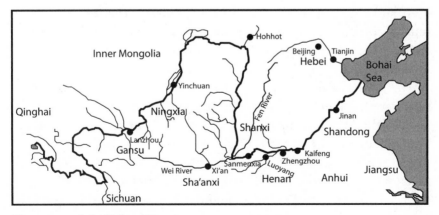

The course of the Yellow River.

It is in the denuded and rugged landscapes of Ningxia, Sha'anxi and Shanxi that the river gets it hue. This region is a vast plateau of loose sandy soil called loess, hundreds of metres thick, blown there from the Gobi desert just to the north in Mongolia. The soil is powdery and virtually free from grit, so that it crumbles to an ochre smear under your fingers. This is China's famous 'yellow earth' (*huangtu*). Loess is easily eroded, and winds blow it in blinding clouds as far east as Beijing. While the capital's now infamous dust storms have been aggravated by desertification in the north-west, they have been apt for centuries to descend and leave everything – houses, trees, animals, people – coated a dirty yellow.

The great river fills with sediment as it carves its course through this landscape, loading the waters with a higher density of solids than is found in almost any other river in the world. From each kilogram of Yellow River water you can extract as much as 300 grams of sediment, making it tantamount to liquid mud. By the time the river turns eastwards again at the threefold meeting of Sha'anxi, Shanxi and Henan, it is a reddish-golden colour.

This sediment gives the Yellow River the Janus nature in which it both nourishes and devastates the nation. The loess-rich water deposits fertile soils in the middle and lower reaches – the North China Plain – where there are great fields of wheat and sorghum, millet, maize and sweet potato, the latter two imported from the New World. Half of China's wheat is grown here, and a third of its maize and cotton. A quarter of the country's population live on these plains, and one estimate maintains that over time more than a trillion people have lived and died here, fed by the rich alluvium. The archaeological remains of agricultural villages have been found from around the eighth millennium BC, which is when millet was first domesticated in China.

The river has been engineered for over two millennia so that it might swell the bounty from farmland. Irrigation here dates back at least to the Warring States period from the fifth century BC, when the feudal system emerged. While anthropologist Jared Diamond's suggestion that agriculture was 'the worst mistake in the history of the human race' shoulders all the burdens of counterfactual histories, there is hardly a better example than the Yellow River to advance his argument. The story of the river basin has been one of interactions between human civilization and nature that constantly raised the stakes while at the same time creating an artificial ecosystem of vast scale and perilous fragility: a landscape almost wholly shaped by human agency, yet nonetheless still massively vulnerable to nature's whims.

For, although most major rivers are prone to flood, the Yellow River valley has suffered from it in a manner both extreme in extent and seemingly intractable in cause. As the river flows east, some of the sediment settles onto the bed, raising it higher. The waters then become increasingly likely to overrun the banks when the flood season arrives with the rains and the melting of snow at the headwaters in summer. To combat flooding, for millennia the Chinese built dykes along the river: huge ramparts of mud, reinforced with sacks of rocks, woven reed mats and clumps of vegetation. But this method of flood control is unsustainable. As the riverbed rose, so did the dykes, until the river itself flowed as if along a semi-natural aqueduct up to fifteen metres above the level of the surrounding floodplain. When a breach in the dykes occurred – and it always did eventually – the result was all the more catastrophic. Kilometres of dykes, having been laboriously built and maintained for years, might be swept away in a matter of hours, and the river water pooled into immense lakes and inland seas. As the flow was diverted, it slowed down and silt was deposited at a greater rate, choking up the old bed and making it extremely hard to return the river to its course.

Yet it was precisely because of the river's fertile sediments that the floodplain was so attractive to farmers, accumulating a rural population at constant hazard of disaster. At the same time, intensive agriculture exacerbated the danger. The demand for cultivable land, as well as for timber to use as fuel and in construction, led to clearance of the forests that once covered the loess plains. The bleak, barren badlands of today, riven by chasms and gorges, are largely a human construct, for the forest cover on the loess plateau is thought to have declined over the past four millennia from more than 50% of the land area to just 8%. Lacking the protection of forest canopy and root systems, the exposed soil is more readily eroded by rain, which not only destroys farmland but also boosts the sediment load in the river, making the problem of silt deposition still more grave.

The effects of land clearance were already felt in the Qin and Han periods two millennia ago, and deforestation was condemned in some ancient texts. The problem worsened considerably during the Tang dynasty, when agriculture intensified to provide food for China's army as the empire expanded its borders and maintained large garrisons against the threat of invasion. It was in Tang times that the river's sediment load earned it the name 'Yellow'.

Increased erosion made the river meander more dramatically, so that farmers could never be sure how long their fields would survive. Moreover, the climate was relatively dry at this time, which increased the pressure on irrigation. That was never done efficiently: fields were simply waterlogged, which meant that mineral salts deeper within the soil were dissolved and carried to the surface. There they accumulated when the water evaporated, producing saline soil with low fertility. (This process of salinization remains a blight of over-irrigation globally today.) Deforested land was sometimes over-farmed and quickly depleted in nutrients, whereupon it was abandoned and yet more land cleared. In this way, what was once farmland became barren ground and eventually desert. With the loess exposed, the river began to meander widely as it cut into and shifted the sandy deposits, creating the other-worldly terrain of ravines and gorges that distinguishes Shanxi today. The American journalist Edgar Snow gave a compelling account of these landforms in the 1930s:

> an infinite variety of queer, embattled shapes – hills, like great castles, like ranges torn by some giant hand, leaving behind the imprint of angry fingers. Fantastic, incredible and sometimes frightening shapes, a world configurated [*sic*] by a mad God – and sometimes a world also of strange surrealist beauty.

Floods of unimaginable proportions have ravaged the Yellow River valley since ancient times. As the Han historian Sima Qian noted, 'Inconceivably great are the benefits and the destruction which water can produce.' Until modern times there were, on average, two breaches of the dykes every three years, although floods have somewhat increased in both frequency and severity over time. The great flood of 1917 elicited a starkly symbolic image: the waters exhumed wooden

Some of the characteristic loess formations of Sha'anxi province.

coffins from their shallow burial mounds and set them floating for many kilometres.

With the Yellow River designated the cradle of the nation's civilization in early Chinese historiography, its moods were linked to the fate of the nation. The massive dam that stands today at Sanmenxia, just after the final eastward turn – one of the earliest modern attempts at flood control on the lower reaches – bears an inscription attributed to the Great Yü, who conquered the Great Flood, that presents something of a glass-half-full perspective: 'When the Yellow River is at Peace, the Nation is at Peace.' The unspoken corollary is that if the river is not at peace, then the nation may rupture too. China's mother river is also China's Sorrow.

There are many other nicknames attesting to the river's unruly nature: the Ungovernable, the Scourge of the Sons of Han. Some calamitous floods redrew the map. When it breaches its banks, the Yellow River might never find its way back between them: the inland sea that results from a major flood can find a new route to the coast. Since 600 BC there have been dozens of such shifts, eight of them classified as 'major', meaning that the outflow into the ocean may be hundreds of kilometres from its earlier location.

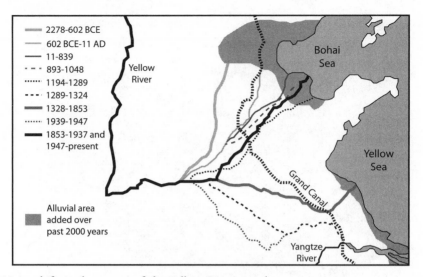

2278-602 BCE	
602 BCE-11 AD	
11-839	
893-1048	
1194-1289	
1289-1324	
1328-1853	
1939-1947	
1853-1937 and 1947-present	

Alluvial area added over past 2000 years

Major shifts in the course of the Yellow River over the ages.

Life on the Yellow River floodplain was not so much precarious as predictably disastrous, and it is hard to imagine how anyone, let alone millions, endured it routinely. Even in the modern era the floods could be terrible beyond imagining: a breach in the autumn of 1887 created a lake 26,000 square kilometres in extent, leaving people stranded on rooftops as the bitter northern winter closed in. Between 1 million and 2.5 million people perished by drowning or subsequently by starvation, through epidemic diseases such as typhoid, or from exposure. The hole in the dykes was not plugged until early 1889.

The problems created by the sediment that the Yellow River acquires on its looping northward detour are so great that, as early as the first century BC, emperors were considering whether to circumvent this diversion entirely – to cut a channel east–west that linked the bends across 300 miles. It is hard to imagine how anyone at that time could have considered it feasible. But the Han engineer Yan Nian, who made this bold proposal, argued that not only would it make the river easier to control by reducing its silt load, but it would also offer a better barrier against the encroachment of the 'Huns' (the Xiong Nu) of Mongolia. The emperor rejected the idea not because it was impractical but because it seemed sacrilege to change the course allegedly designated by the Great Yü, who solved the Flood by carving out new channels for China's rivers. Yü, the emperor declared, had acted with 'divine perspicacity . . . for the benefit of ten thousand generations'.

The Long River

One seems to have little choice but to retain the outmoded name for the Yangtze when discussing it in English; the modern Pinyin transliteration Yangzi feels somehow pedantically perverse. The name is in any event only a local one, derived from the ancient and now mostly forgotten fiefdom of Yang and strictly applying only to the last 300 kilometres. This was the entire 'Yangtze' to the first Western travellers, since they rarely got much further upriver.

The Chinese people do not use those names. There are local names for each stretch of the river, but the full channel, cutting the country in half geographically, climatically and culturally, is simply the Chang Jiang (长江), the Long River. It is the longest in all of China, 6,380 kilometres from the source in a glacier lake to the great delta on the coast beyond Shanghai, where the alluvium pushes out into the sea and adds steadily to China's vast surface area.

'A China without such an immense torrent at its heart is almost impossible to contemplate', says the writer Simon Winchester. Even this understates the matter. Without the Yangtze, China would not be the nation it is today. Time and again, the river has determined the nation's fate, whether that is by presenting a barrier to barbarian conquest, or a transport network, or a conduit for foreign invasion, or a source of fertility, flood and revolutionary fervour. Many pivotal battles in Chinese history took place on the middle reaches. The Yangtze cliffs provide the backdrop to the classic *Sanguo yanyi* (*Romance of the Three Kingdoms*) from the early Ming period, one of Mao Zedong's favourite books, in which the river hosts allegedly the biggest naval battle in history. The Yangtze was the artery of conquest and dominance when the British gunships humiliated the Qing emperor in the mid-nineteenth century, and again when the Japanese invaded in the 1930s: steadily pushing upriver from Shanghai to Nanjing and then Wuhan, they forced Chiang Kai-shek's Nationalists to relocate the government right back beyond the Three Gorges in Chongqing.

China is cloven in two by the Long River, and the two halves could seem like separate nations: the north cold and dry, the south hot and wet. In the north you eat wheat noodles; in the south, rice. Northerners, it is said, are tall and haughty, whether eastern Manchurian stock or Islamic Uyghurs to the west. The southerners, in contrast, are earthy, pragmatic, always on the make, a patchwork of minority races and mutually incomprehensible dialects. That division – decreed by nature, patrolled by the Yangtze – establishes the defining tension within the nation, in which the question is how unity can persist in the face of such a disparity of the most fundamental resource, water. Such stereotypical polarities do scant justice to the bewildering variety of China, of course, but they serve as crude shorthand for the contrasts that you find once you cross the Yangtze.

For the mixed blessing of the Yangtze, with a valley rich in farmland yet also suffering enormous floods, the Chinese again credit the Great Yü. From its source the river flows south, parallel to the Mekong and the Salween as the three great torrents plunge down gorges like giant sword-strikes through the mountains of Tibet and Yunnan, heading out of China in short order. But then at a place called Shigu in Yunnan, the Yangtze leaves the trio as it takes a remarkable bend, seeming almost to bounce off a modest little mountain called Yun Ling (Cloud Mountain) to execute an abrupt about-turn and then find its way east instead. Yü is said to have set down Cloud Mountain; no other legendary figure could be entrusted with the task of defining the course of China's central artery.

In Sichuan province the Yangtze is swollen by tributary rivers running south from the Qinghai highlands, in particular the Min,

The great bend in the Yangtze at Shigu.

Yalong, Dadu, Fu and Jialing.* It has descended 90% of its source
altitude even before it passes through the bustling, steep-laned citadel
of Chongqing, the epitome of China's frenetic enterprise (and, some
say, the birthplace of Yü). Then it wanders through Hubei to the
vigorous trading port of Wuhan, a conglomeration of the former
cities of Wuchang, Hankou and Hanyang, where the Yangtze intersects
the Han River. This political and economic hub of central China
was the origin of the Wuchang uprising that ended the Qing empire
in the early twentieth century and gave birth to Sun Yat-sen's Republic.
The river then courses majestically across the eastern plains of Anhui
and Jiangsu, through the southern metropolis of Nanjing with its long
tally of bitter memories, before spilling out past Shanghai into the
East China Sea.

While the Yellow River was commandeered in the early days of
the Republic for the active construction of a national identity-myth, it
is the Yangtze that today defines China's self-image. To travel the river
from source to sea is to sail down the currents of history. In the upper
reaches of the Qinghai plateau one can find a way of life, often close
to destitution, that has changed little for centuries (apart from the
ubiquitous cheap mobile phones), while Shanghai, that promiscuous
old harlot on the Huangpu tributary in the great estuary, exemplifies
the brash, confident, almost unstoppable spirit of modern China.
Along the route one will find some of the country's most spectacular
scenery, its most astonishing and controversial feats of hydraulic engineering, its greatest lakes, ancient cities like Jingzhou, Yangzhou and
Nanjing, bleak and despondent industrial centres, dynamic river ports
still bearing the traces of colonialism, sites of momentous struggles.
There are rice paddies knee-deep in river water, temples and pavilions
where poets sat and wove watery metaphors, there are mythical
mountains and filth-belching factories. Even the bustle of commerce
that has always intruded on the navigable reaches does not wholly
dim the beauty that the Song administrator Lu You rhapsodized about
in the twelfth century:

---

* It is often said that Sichuan derives its name from four of these tributaries, since
the name means simply 'Four Rivers'. But that isn't quite right: *chuan* (川) does mean
'river', but the provincial name is shortened from *chuanxia silu* (川峡四路), the 'four
circuits of rivers and gorges', referring to the Song dynasty administrative division
of a 'circuit' (*lu*).

The Caiyuanba Bridge over the Yangtze in Chongqing.

All the solitary hills in the midst of the River, such as Golden Mount, Jiao's Hill, Fallen Star and the like, are famed throughout the world, but for dizzy heights and elegant beauty none can match the Lesser Lone Hill. Seen from a dozen or so miles away, its bluish peak rising abruptly all alone, its top touching the high heavens, it already seems beyond compare with other hills; and the nearer you approach the more elegant it is. In winter or summer, in clear skies or rain, it presents a myriad different moods. It is truly a marvel of Creation.

In some ways, life along the Yangtze has changed little since Lu You described it: his 'crowds of young lads along the water's edge selling caltrops and lotus roots' are still there, although they might equally be selling fake branded goods and pirated DVDs. In the lower reaches of the Yangtze, entire villages once floated on the waters. Lu You describes them:

As we tacked along the Great River we came across a wooden raft of one hundred feet or more in breadth and over five hundred feet long. There were thirty or forty households on it, with a full complement of wives and children, chickens and dogs, pestles and mortars. It was criss-crossed with paths and alleys, and even had a shrine to a deity. I've never set eyes on such a thing before, but the boatman said that this was still one of the small ones. The large ones have soil spread on the raft for vegetable allotments, and some have wine shops built on them.

These floating villages were common even up to the middle of the twentieth century.

However much writers and artists might romanticize it, the Yangtze most aptly symbolizes the Chinese nation insofar as it serves as a trade thoroughfare. More than three-quarters of China's rice is now produced in the paddy fields of the lower reaches, and for centuries transport of this grain to the power centres of the north was one of the emperors' key priorities. The colonialist struggles of the nineteenth century focused on control of the Yangtze ports, and the river has long been engineered to push westward the limits of navigability. Even in the Han dynasty, Sima Qian labelled the lower Yangtze the 'land of fish and rice', where the people were so easily fed that they became lazy. The river once yielded half of China's fish, although that proportion has now declined because of pollution and near depletion of stocks (commercial fishing is now highly regulated). The lower reaches were also a key region of silk production, which found its way into the wide world not by caravan along the Silk Road in the north but eastwards by ship over the East China Sea.

The Yangtze is literally shaping China. It carries along 500 million tonnes of alluvial silt each year – nowhere near the heavy load of the Yellow River, but enough to push back the coastline by about a kilometre and a half every seventy years as the deposits settle in the delta, leaving the cities of Ningbo and Hangzhou tens of kilometres from the sea when once they were ports. The intrepid English Victorian explorer Isabella Bird offered a quaintly precise estimate of the annual sediment load – 182,044,996 cubic metres – but also a vivid description of the material that turns the Yellow Sea yellow: 'The rich wash of scarcely explored Central Asian mountain ranges, the red loam of the "Red Basin" of Sze Chuan, and the grey and yellow alluvium of the Central Provinces of China.' The great sandbar in the mouth of the Yangtze was dredged in 1905 by the European colonialists who prevailed over the fatally weakened Qing empire, literally opening the way for Shanghai to become the playground of Western merchants until disaster fell upon it (and upon all of China) when the Japanese invaded.

The Yangtze is quite capable of arranging disasters of its own making. Its floods are no less devastating than those of the Yellow River, nor any less regular. Those in 1931, coming after a summer of particularly heavy rains, are thought to have killed up to 4 million people and inundated an area almost the size of Great Britain. A description of another massive flood in 1887 by the Englishman William Percival paints a terrifying picture of what the locals faced. A great wave could be seen coming down the river, he writes,

> carrying with it numbers of junks, boats, houses, trees, cattle, and I should be afraid to say how many human beings, all mixed up in the most inextricable confusion . . . Houses floated past with people clinging to them, some hanging on to the branches of trees, while scores of corpses and the bodies of cattle seemed all over the river. Everything not drowned, everything living, both human and animal, were yelling, roaring, and screeching. All this, combined with the grating and crashing of houses, the sullen rush of water, the howling of the wind, and the swish, swish of the blinding rain, made such a pandemonium that I hope never to see again.

While flooding on the Yellow River seems to have been brought somewhat under control in recent decades, the Yangtze remains deadly. There were floods throughout the 1980s and 90s; a particularly serious inundation in 1991 affected 230 million people and forced the Chinese government to swallow its pride and ask for international relief (it received rather little). The Yangtze struck again in 1998, when floods killed around 3,500 people and caused damage of the order of $20 billion. Human activity seems to have worsened the risks. Reclamation of lakes along the river for farmland has removed natural flood reservoirs. And erosion in the upper Yangtze valley, which has raised the riverbed in the middle reaches by siltation, has been exacerbated by extensive deforestation of western Sichuan since the 1950s for timber, prompting speculation that the Yangtze might come to rival the Yellow in its high load of silt.

Pearl S. Buck, the Nobel laureate chronicler of early twentieth-century China, attested how, during one flood, 'I stood on Purple Mountain in Nanking [Nanjing], many miles distant from the river, and it was a great island, and lapping at its base, fifty feet deep over farmhouses and fields, were yellow Yangtze waves.' 'There is no other river', she added, 'to equal it for beauty and cruelty.'

Petrified witches

The most celebrated stretch of the Yangtze is in the middle reaches between Chongqing and Yichang. Here the waters surge through the Three Gorges: Xiling, Wu and Qutang. Tightly focused into a fast-flowing torrent by the precipitous limestone slopes and cliffs that stretch, with intermittent breaks, for almost 200 kilometres, the river here is legendary for its beauty and its perils. The plunge from clifftop to turbid water is a little shorter today than it was in the past, and the waterway a little wider, for the river has become a long, narrow reservoir confined behind the mighty wall of the Three Gorges Dam.

It is this stretch of the Yangtze that is most heavily mythologized: every rock and mountain, it seems, warrants a name and a story that explains its presence and shape. In general the stories tell of some great battle that pitted benevolent gods, demigods or heroes against raging, chaotic nature, often personified (like the river itself) as a dragon. They are tales of taming and the creation of order from the wildness of the waters and the elements. Like many myths, they enabled the tellers to dream of what might be possible, even if it was beyond their capabilities.

Presiding over the entrance to the Wu Gorge – so steep that direct sunlight reaches the water only for a short time even in summer – are twelve peaks, said to be the petrified bodies of the witch-goddess (or fairy) Yao Ji and her sisters, who reputedly overcame twelve mighty river-dragons that were bringing misery to the local people. The slain dragons, however, became great boulders, damming the Yangtze and creating a terrible flood.

Then the flood-master Yü arrived to put things in order. At first he was dumbfounded by the situation, but Yao Ji calmed his despairing sighs and ordered lightning and thunder to break apart the rocks and drain the water. Near the region of Wushan (Wu Mountain), Yao Ji and her sister spirits helped Yü to construct a tunnel through the mountain. The Qutang Gorge was an even greater challenge, for a jade dragon who dwelt in a small tributary called the Daixi at the eastern end of the gorge was causing mischief. He became so angry after getting lost visiting relatives (even dragons in China observe their

The Three Gorges region on the Yangtze River.

familial obligations) that he threw his body at the mountainside, creating a landslide and flood. Yao Ji chained the dragon to a pillar so that Yü could cut off its head. The flood-master then opened a new channel for the river. Two great black rocks in the Suokai (Unlocked Gates) Gorge downriver are now named the Binding Dragon Pillar (Suolong Zhu) and Beheading Dragon Platform (Zhanglong Tai).

Yao Ji and her assistants stayed in the region to help the people: blessing their harvests, tending their sick, and watching over riverboats. But they found the Wu Gorge so beautiful that, standing every day on the cliffs gazing at the river, all twelve of them were turned to stones that now preside almost 300 metres above the water. Perhaps, some say, they still act as sentinels for the boatmen braving the rapids. A temple to Yao Ji and Yü on Gaoqiu Mountain near Wushan commemorates their efforts, and on 'Goddess Day' offerings are made at her shrines.

Yao Ji embodied the old association of goddesses with water: she was said to control the rain over the gorge. As a Tang dynasty monk put it:

> Witch mountain is high
> The witch is uncanny,
> As rain, she brings the sunset,
> Oh! As cloud, she brings the dawn.

Goddess (Shennu) Peak, the most prominent of the twelve peaks on Wu Gorge, said to be the petrified body of the spirit Yao Ji.

The scenery was irresistible. On an outcrop near Qutang – now an island because of the raised water level behind the Three Gorges Dam – stands an ancient temple complex called Baidicheng (White Emperor City), which the Tang poet Li Bai eulogized ('engulfed by vibrant clouds') as he sailed down the Yangtze. Li Bai's friend Du Fu, who is widely considered China's greatest poet, spent two or three years in the nearby city of Kuizhou (now Fengjie), and would come often to Baidicheng and sit alone up on the mountain – he composed a quarter of his entire *oeuvre* here in those few years. In the Song period Lu You travelled up the Yangtze to Kuizhou and was awed by these peaks and cliffs, 'some vying one with another, others soaring solitary'. 'I cannot fully describe their wondrous strangeness', he averred.

Westerners arriving in the nineteenth century had seen nothing like this, even in the Alps or Scandinavia. Isabella Bird gives a vivid account of her passage through Yichang Gorge:

We were then on what looked like a mountain lake. No outlet was visible; mountains rose clear and grim against a grey sky. Snowflakes fell sparsely and gently in a perfectly still atmosphere. We cast off from the shore; the oars were plied to a wild chorus; what looked like a cleft in the rock

appeared, and making an abrupt turn around a high rocky point in all
the thrill of novelty and expectation, we were in the Ichang Gorge, the
first and one of the grandest of those gigantic clefts through which
the Great River, at times a mile in breadth, there compressed into a limit
of from 400 to 150 yards, has carved a passage through the mountains.

There are caves here where ancient fossils have been found that
rewrote the standard accounts of human origins (see page 51). They
are full too of fossil animal bones, once harvested by peasants and
sold for medicinal use as dragons' bones and teeth. Perhaps the most
wondrous and strange sight in the gorges are the 'hanging coffins'
placed on wooden pegs inserted into the sheer cliff face by an ancient,
legendary minority called the Ba. Some of the coffins, made from
waterproof, durable *nanmu* wood, are around 2,000 years old. They
can be found all along the gorges and some of its tributaries. How
they were elevated is still unclear, although it seems likely that
ropes were used to haul them up. Why they are there is another
matter. Were they a perpetual reminder to revere your ancestors?
(There's nothing more guaranteed to do that than having them literally looming overhead.) Did they bear the spirits of the departed to
the land of the dead, or back into nature, in a kind of celestial boat?
Did they ward off demons? Making them so inaccessible has certainly
been effective in warding off grave-plunderers hoping that the rumours
of great riches inside the caskets are true, not to mention the worst
destructive excesses of the Cultural Revolution's Red Guards (who
nevertheless managed a little vandalism). The few coffins that have
been examined since archaeologists first reached them in 1971 contain
bronze and pottery artefacts, along with the skeletons of the dead.

Until the late nineteenth century, Yichang at the eastern mouth of
the Xiling Gorge was the last stop navigable by large vessels, and as
a result it became a city of more consequence than its rather down-
at-heel appearance today would suggest. British companies set up
trading posts there, and they levied duties on all goods that passed
through, of which opium was one of the most lucrative. To get any
further upriver was a perilous affair. Li Bai claimed that his hair was
turned white by the terror of passing through Wu Gorge, while White
Bone Pagoda on the notorious Xintan rapids in Xiling Gorge was said
to house the bones of boatmen drowned there. Lu You records that

there were efforts even in Song times to clear away the deadly shoals lurking just beneath the water's surging surface – but that locals, unwilling to forfeit the plunder they enjoyed from shipwrecks, would do their best to hinder the efforts, bribing quarrymen into claiming that the rocks were unmovable. The worst of these have now been blasted away with dynamite. From the mid-nineteenth century a life-boat service called the Red Boats operated along this stretch, financed initially by a wealthy merchant. By the early twentieth century, a fleet of around fifty Red Boats saved more than 1,000 lives every year, as well as acting as de facto river police.

To move upstream, ships would employ (or, in older times, simply compel) teams of men called trackers to pull them against the current. It was an inhuman feat, in all respects. As the historian Lyman van Slyke writes:

Lukan Gorge, as drawn for British explorer Thomas Blakiston's *Five Months of the Yang-Tsze* (1862). Isabella Bird photographed this gorge three decades later.

Tracking a 120-ton Mayangtzu [a kind of junk] through the worst rapids in the Three Gorges, against some of the most difficult currents in the world, is a feat so apparently impossible that the Western mind wonders how it could have been conceived in the first place, let alone accomplished.

But it was. A team might contain up to a hundred men, all heaving on a cable as they trod along narrow paths sometimes hewn into the rock face of a gorge and dangerously slippery with spray. The cable was attached to the boat's hull and looped around the mast; it was traditionally made of plaited bamboo, which has half the strength of mild steel and the added virtue of becoming stronger when wet and under tension. Tracking was established by the Tang era, when Li Bai mentions it in a poem. Lu You attests that no one tried to traverse the gorges by sail, and says that the trackers' tow ropes were 'as thick as a man's arm' and 300 metres long, which sounds like a gross exaggeration until one sees that Isabella Bird's account in the late nineteenth century corroborates his figure. She says that these 'monstrous' coils of rope, lodged on the roofs of boats, lasted only for a single voyage, which was not surprising when one saw the deep scores that the ropes abraded in the granite rock. The Victorian navigator and entrepreneur Archibald Little, who founded a steamship company on the upper Yangtze in 1887,* describes the trackers' business:

A big junk of 150 tons carries a crew of over 100 men, viz. seventy or eighty trackers, whose movements are directed by the beat of a drum, the drummer remaining on board under the direction of the helmsman; a dozen or twenty men left on board to pole, and fend off the boulders and rocky points as she scraped along, and also to work the gigantic bow sweep formed of a young fir-tree. Another half-dozen of the crew are told off to skip over the rocks like cats, and free the towline from the rocky corners in which it is perpetually catching; besides a staff

---

* Little's business, and later that of his captain Cornell Plant, pioneered the commercial navigation of the upper Yangtze between Yichang and Chongqing – as Little explained, 'the Yang-tse is not only the main, but the sole road of intercommunication between the east and west of this vast empire'. Thanks to his efforts in opening up that road, Plant was made a River Inspector for the Chinese Imperial Maritime Service, and was eventually honoured and commemorated by the government, which built a house for him and his wife overlooking the rapids.

of three or four special swimmers called '*tai-wan-ti*' or water-trackers, who run along, naked as Adam before the fall, and may be seen squatting on their haunches on rocks ahead, like so many vultures, prepared to jump into the water at a moment's notice and free the towline, should it catch on a rock inaccessible from the shore.

Tracking was a terribly hard business. These men, often working naked, were some of the original 'coolies': the anglicized word for any labourers who bore heavy loads, derived from *ku li*, 'bitter strength'. In older times they were whipped by 'gangers' (as the British called them) to 'encourage' the exertions. Sometimes, it was said, the contract for hiring a tracker would include a clause guaranteeing to supply his coffin. Isabella Bird testified that 'On every man almost are to be seen cuts, bruises, wounds, weals, bad sores from cutaneous disease, and a general look of inferior rice.' Tracking came to an end only in the 1950s, when the worst of the rapids were blasted away with dynamite.

The Yangtze was plied by merchants of all kinds. The richest of them traded salt: every peasant in China, lacking a meat-rich diet, needed that. (Soy sauce was invented to eke out this costly commodity.) Salt trade was controlled by the state, but private producers and sellers could buy licences to sell in authorized districts, and grew fat, perhaps decadent, on the proceeds. Other traders made their fortunes with grain, tea, vegetable oils, medicinal herbs, hides and furs, cotton, beans, hemp, timber, bamboo or coal.

Yet mercantile life was precarious, for transporting anything along the middle Yangtze was fraught with hazard. When Archibald Little set up his business, it was said to be faster and cheaper to get a tonne of cargo from London to Hankou than from Hankou through the gorges to Chongqing. Cornell Plant – who piloted Little's 55-metre ship the *Pioneer* from Yichang to Chongqing in 1900 before setting up his own steamship business eight years later – indicated that one junk in ten was badly damaged every time that journey was attempted, and that one in twenty was wrecked beyond repair. River captains hired out their vessels and services at considerable risk, and extended no mercy to the trackers, as Plant relates:

The *laopan* [*laoban*: boss], or skipper, usually perches on top of the deck house, from which point he watches moves. When any hitch occurs he rises to the occasion, stamping, cursing, and raving – often returned to him by the gangers, who go one better – until the skipper's mate, Mrs Laopan, cannot stand it any longer. She comes on deck, and with a few well-directed vexatious epithets ends the row. Even a Thames bargeman would be dumbfounded before Mrs Laopan.

Sailing up the Yangtze could demand great skill and courage even in modern times.* As the captain of a large passenger ship that made the journey between Shanghai and Chongqing in the 1980s put it:

> I think even the most restless person can become patient after sailing so many years on the Yangtze. But we aren't like those seamen whose horizons are broader. We sail a narrow course and we must be very prudent and cautious.

He could be voicing a metaphor for China's leaders.

Poetry of the lakes

As the Yangtze levels and broadens along the borders of Hubei, Hunan and Jiangxi provinces, its waters feed into China's two largest lakes, Dongting and Poyang. These serve as natural flood basins: they swell and shrink with the river's seasonal flow. For Lu You, Poyang was 'a boundless expanse wherever you looked'; Li Bai called it a 'celestial looking glass'. Dongting, normally covering around 3,000 square kilometres, can grow to almost seven times that size in the flood season. This regular pulse dictated the rhythms of life on the lake shore. In the dry season of winter and early spring, farmers went onto the exposed sandy marshland to plant rice paddies; as the waters began to rise, they moved back outwards. The provinces that Dongting straddles owe their names to the lake: Hunan is 'South [of the] Lake', Hubei 'North [of the] Lake'.

This immense reservoir has shrunk significantly as land has been reclaimed for farming. Once Dongting exceeded Poyang in size, but

---

* Today the risks for passengers on the Yangtze cruises are more from tour companies' attempts to cut costs by using poorly maintained ships and inexperienced crews. But the elements can still play a role: when the *Oriental Star* capsized and sank in June 2015, killing nearly all of the 454 passengers and crew, extreme weather – a tornado and torrential downpour – exposed its poor safety measures.

its area declined by almost half over the course of the twentieth century. At the driest point in the annual cycle it becomes little more than a network of muddy estuaries. Dongting has been a source of dispute between Hubei and Hunan since the 1950s, with the northerners protesting that the southerners were reclaiming so much land from the lake that it could no longer act as a flood reservoir, leaving them at risk. The Yangtze River Water Resources Commission tried to find a clumsy compromise by constructing floodgates at the lake entrance that could be closed most of the time to create farmland but which would be opened if a risk of flooding made that necessary. What this solution didn't recognize is that the people of Hunan wouldn't just farm the land in temporary fashion – they would populate it with houses, roads and schools, making it quite out of the question to open the gates to avert a flood.

Before factories began to colonize their shores, the lakes were sites of great beauty – and they have not lost all their charms today, despite the havoc being wreaked with their ecosystems. Poets and painters came to take in the marshes and reeds, the placid waters reflecting the moonlight, the fishing vessels going about their resolute and solitary business. In the Tang era, Dongting inspired almost its own genre of poetry. The poet Zhang Yue was awed by its scale:

> Vast, vast: lose all sense of direction,
> Turbid, turbid: like congealed *yin*;
> Clouds and mountains emerge and disappear,
> Heaven and earth float and sink.

Li Bai lived for a time on the lake's shores, and like many poets he looked out over the lake from Yueyang Tower, built as a military fortification in the Three Kingdoms period:*

> From the tower, a prospect of all Yueyang,
> Different from rivers, Dongting is open;
> Geese lead sorrowful hearts away,
> In mountain jaws a fine moon comes.

---

* The original tower burnt down in 1079; the building that stands in its place today is a more recent reconstruction.

Swimming the Yangtze

As the invading Japanese forces stood poised to seize Wuhan in 1938, the poet Zou Difan invoked the Yangtze as a spirit of resistance, a reminder that China had survived hard times in the past and would do so now:

> Ah, river –
> I love your robust strength
> That carries you thousands upon thousands of miles.
> No dyke can stop your surging billows
> From irrigating the land and feeding the people along your course
> For thousands of years without end.

For Mao Zedong, however, the fact that no dyke could contain it made the Long River a dragon to be conquered, demonstrating that the nation was now in the hands not of nature but of the party and its leaders. In part the taming involved eye-catching technological mastery: the first major bridge over the main stretch of the river, inaugurated at Wuhan in 1957, was a double-decker structure carrying roadway and rail line. Mao had anticipated this glorious achievement in a poem of the previous year, at which point the structure was already under construction: 'A bridge will fly to span the north and south, turning a deep chasm into a thoroughfare.'

But that poem marked a still more symbolic demonstration of personal authority over China's mythical waterway. The People's Republic was established in 1949, but by the mid-1950s the optimism of the Communist state's birth was giving way to the trials of reality – including a catastrophic Yangtze flood in 1954. Mao needed a grand gesture to assert his leadership. In the summer of 1956, before the gaze of the propaganda cameras, he stripped on the shore of the Long River at Wuhan and thrust out into the treacherous current. The iconic photographs of this event show Mao apparently isolated in the water, but in fact he was accompanied by an entourage of guards and officials. They all had good reason to be anxious. During a practice swim of the Xiang River in Hunan, a guard was bitten by a water snake. And the two soldiers who had swum the Yangtze before Mao to ensure it could be done had encountered currents and water-borne parasites sufficient to persuade them that it was not safe. But they did not dare tell their leader that.

The opening of the Wuhan Yangtze bridge in October 1957.

The feat demonstrated more than just Mao's physical strength and endurance; it was a show of mastery over nature, and illustrated yet another lesson to learn from water. As Mao wrote in 1957:

> To draw an analogy, the people are like the water and the leaders at all levels are like the swimmers. You mustn't leave the water. You must go with [the flow of] the water, you mustn't go against the water.

Facing down the Yangtze also offered Mao a potent metaphor for China's might beyond its borders. 'The Yangtze is a big river, people say', he attested:

> It is big, but not frightening. Is imperialist America big? We challenged it; nothing happened. So there are things in this world that are big but not frightening.

The feat bore repeating whenever the occasion demanded. When rumours circulated in 1966 that the seventy-two-year-old Mao was in ill health, he silenced them with another media-staged swim in the

Yangtze. The Chinese newspapers obligingly conveyed the message. 'Our respected and beloved leader Chairman Mao is in such wonderful health', they declared. 'This is the greatest happiness for . . . revolutionary people throughout the world.'

Mao's mastery of water through swimming became a patriotic example that citizens had a duty to emulate. Propaganda posters in the late Cultural Revolution showed healthy young children swimming, and proclaimed that everyone should learn to do so. Like building the Communist Utopia, it required perseverance and discipline. 'If you would just work at it an hour every day, without fail, go today, tomorrow, for a hundred days, I guarantee you would learn to swim', Mao wrote. His first swim across the Yangtze is still commemorated today in Wuhan and in Changsha in Hunan, where at midsummer people compete in races in the murky water, as dangerous now for its pollution as for its parasites or its currents. Some bathers carry posters of the former leader into the waters, or red banners displaying motivational slogans. In Wuhan some brave souls even venture into the waters in December to mark Mao's

A 1967 poster celebrating Mao Zedong's swimming of the Yangtze the previous year. The caption reads 'Commemorate the first anniversary of Chairman Mao's swim over the Yangtze – Follow Chairman Mao in moving forward in wind and waves!'

birthday. These aquatic feats retain their association with political legitimacy, although the motivation now may be tellingly changed. When in July 2014 the Mayor of Guangzhou donned the obligatory red swimming cap on the anniversary of Mao's 1966 dip in the Yangtze and waded into the waters of the Pearl River, he was hoping to convince watchers that the river waters were unpolluted. In China, symbolic gestures that make use of water have their own momentum, making them ripe for repurposing. The language of water is well understood. It has been spoken in China since the earliest times.

2 Out of the Water

The Myths and Origins of Ancient China

> In the time of Yao the waters reversed their course and over-
> flowed the middle kingdoms so that snakes and dragons dwelt
> there . . . [Yao] had Yü put it in order.
>
> *Mencius*

Lu You can't have been happy about the appointment he received in 1169. This minor official of the Southern Song dynasty, whom we encountered in the previous chapter, was assigned to Kui prefecture (or Kuizhou) on the Yangtze. In modern times it became the grimy but historic town of Fengjie, which was demolished in the early 2000s before being submerged beneath the Three Gorges Dam reservoir. The place was about 2,900 kilometres from Lu's home in Shaoxing, south-west of modern Shanghai in Zhejiang province, and he wouldn't have relished the prospect of getting there. He would have to travel with his family by boat all the way up the Grand Canal to the Long River, and then brave the arduous and dangerous journey of about 1,600 kilometres upstream to a destination regarded as a backwater populated by barely human savages.

Born into a well-bred family from the region south of the Southern Song capital of Hangzhou, Lu passed the civil service examinations that qualified gentlemen for government roles but subsequently failed to attain any position of importance. In 1166 he quit his official career entirely, only to have to resume it because he needed the money. That meant accepting whichever post came his way.

Lu had aspirations beyond the humble station he was allotted. Steeped in the Chinese classics and poets, he hoped to become recognized for his own intellectual and artistic assets. In his later life these aims were realized: while never amounting to much in the governmental structure, Lu became celebrated as a poet and histor-ian. When he set out for Kui, however, he lacked any reputation to speak of, and perhaps partly for this reason he kept a conspicuously erudite diary of his 157-day journey, filled with quotations from famous writers, historical asides and comments on the local customs, geography and nature. As well as demonstrating his prodigious learning, this record might remind the Song court that he was after all a man of distinction, even if his current fate was to while away the days (seven long years, as it happened) far from the centres of power and influence.

Lu's travel diary is one of the earliest of its kind, and seems far less embellished and fanciful than the accounts of travels in the Far East by Europeans a century or so later, notably Marco Polo and the pseudonymous John Mandeville. That tradition of the Oriental travelogue, begun by medieval Western travellers, was sustained by Jesuit monks and missionaries, and then by intrepid Victorian explorers such as Thomas Blakiston and Isabella Bird. For those writers, China was a place of exotic wonder and mystery, as well as squalor and discomfort. They marvel at the sights along the Yangtze and often admire the ingenuity and resourcefulness of the Chinese, while sniffing with disapproval at the sometimes dirty and dilapidated cities to which they assign erratic romanized approximations of the Chinese names. In the differences between the travel diaries of Marco Polo, Blakiston or Bird and that of Lu You we can see what the visitors missed. It's inevitable that much is invisible or mystifying to the foreigners; but as Lu sails past rocks and promontories, temples and shrines, he is evidently passing through a landscape that even in that distant age had already been thoroughly humanized and mythologized. These hills and mountains, gorges and rivers seem all to have names and stories attached to them: they embody a narrative of the land itself, a tale of dragons and warriors, of dynasties and wars, dreams and horrors.

What seemed a wild, pristine and alien terrain to Bird and Blakiston was for Lu a world defined through landmarks of almost cosy familiarity, each of them seeming (when their names are translated) to encode some stirring tale or legend. 'We passed Lion Promontory, which is also called Buddha's Finger Promontory', Lu attests – or 'We then went to the Monastery of the Ascendant Dragon in Grand Tranquillity at East Forest. The monastery directly faces the Incense Brazier Peak.' Scattered Flowers Gorge, Cloud Disperser Peak, Long Wind Sands, Pungent Island, Compassionate Mother Promontory . . . the names make you wish Lu You would tell us more, although of course his intended audience didn't need a primer. Along the Yangtze (a very special location, granted), twelfth-century China was a place alive with legend, a landscape already engraved by the long memory of its history.

We need to know this before entering into Chinese mythology. A common view holds that myth grows from the need to organize and articulate what seems universal in a culture's human experience: as the philosopher Hans Blumenberg has put it, the stories 'selected for repetition' as myth are those 'that seize attention and help people cope with their world'. If this is so, myth can reveal what it is that any particular culture found most challenging to their way of life. Chinese myths, infused with a sense of place and with the perils of natural phenomena, while at the same time preoccupied with questions of authority, administration and succession, show us what it meant to forge and sustain an orderly nation on such a scale in the face of the most extraordinary demographic, political, geographical and meteorological obstacles. Whereas the Greek and Norse gods enchant and amuse us with their very humanity – with their characterful exploits, rages, jealousies and caprices – the Chinese deities are often remote administrators who impose order by superhuman feats of endurance and will.

What these emperor-gods must defeat is in the end no different to what the Communist Party of China still fears today: a descent into chaos and dissolution. In Chinese myth that destructive force often took the very tangible form of water.

From gods to men

Chinese myths don't form a coherent corpus, but which myths ever do? To the extent that the stories exist in several versions that overlap and conflict, they are no different from the myths of ancient Greece.

Yet in China these ambiguities and elisions reflect not so much the disorganizing effects of time as the agendas of those who have retold the tales. In a culture where ancestry and precedent are afforded such importance, myth provided a powerful schema for justifying political decisions and establishing social norms. While cause and effect are hard to disentangle, these features are exacerbated by the fragmentary nature of the Chinese mythical landscape, where, according to the historian Wu Kuo-Chen, anecdotes are 'abruptly told without any sense of organization or artistry, whose characters seem to have entered from nowhere, and then exited into the same nothingness'. It is often unclear if these characters are human or divine, men or monsters, indeed even whether the same name has become affixed to different individuals. Probably one should best approach Chinese myth and legend as a cast of players who were marshalled into various configurations, like pieces on a chessboard, for the convenience of the teller.

The classical creation myths of China are in fact not very old. They are generally considered to date from the time known as the Eastern Zhou dynasty, from around the eighth to the third century BC. In the version that eventually became orthodox, the world was formed within a primeval vapour (the archetypal chaos) from the dying body of a semi-divine giant called Pan Gu. His blood and semen became the waters, and his muscles and veins the water's arterial courses. This world was regarded as a square of land resting on four great pillars, under the dome of the sky from which Pan Gu's sweat fell as rain. The giant's eyes became the sun and moon, while humans had an undistinguished origin in the worms or mites that infested his corpse.

The origin of humans is alternatively described in a still older myth: they are made by a goddess variously transliterated as Nü Gua or Nu Wa, who makes them from yellow earth and mud to ease her loneliness. Nü Gua was wedded to Fu Xi (or Pao Xi), a kind of Chinese Prometheus attributed with inventing hunting, cooking and human technology generally, along with the system of divination in the *I Ching* (*Yi jing: The Book of Changes*). In some myths Nü Gua and Fu Xi are sister and brother who survive a great flood and are given divine sanction to procreate incestuously and repopulate the earth. Some ancient texts list them among the Three Sovereigns: godlike beings who used their magical powers to foster skills and knowledge among the earliest members of the human race, and thereby to enable civilization to emerge.

Fu Xi derived the eight trigrams of the *I Ching* from the 'River Scheme' (*hetu*), which was borne to him out of the Yellow River by a dragon-horse. The scheme is a cryptic square of symbols, allegedly a kind of diagram or 'map' of the Yellow River itself. At much the same time – the exact chronology is ambiguous – the Chinese people offered sacrifices to the god of the Luo River, a tributary of the Yellow River, to appease his anger, whereupon a magical turtle emerged from the river with another cryptogram imprinted on its shell. Called the Luo Script (*luo shu*), this is a more conventional numerological magic square, represented symbolically, which supplies the basis for the geomantic scheme of *feng shui*. Thus the Yellow River was the source of the mystical organizing systems on which Chinese civilization was founded.

The cosmogenic era of the Three Sovereigns was followed by the period of the Five Emperors, the first great sage-kings who ruled China. Their identities vary in different sources, but all agree that the first of them is the Yellow Emperor (*Huangdi*), who became something of a cult figure during the Warring States period of the fifth to the third century BC. The Yellow Emperor was said to have founded Chinese culture along the middle reaches of the Yellow River – the region later called Wei – in a period traditionally dated, with unlikely precision, to between 2697 and 2597 BC. This remote era is often considered historically to mark the transition from a nomadic to an agrarian culture: the Yellow Emperor's predecessor, Shennong (generally listed as one of the Three Sovereigns), is the 'Divine Farmer' who taught the Chinese people how to cultivate crops. The Yellow Emperor's principal wife, Luozu, meanwhile, is said to have begun silk cultivation.

The Yellow Emperor is, among other things, a water deity: like the godlike Jade Emperor before him (see page 53), he commands the rain and clouds. During the period historically attributed to his rule, the predominant natural hazard on the North China Plain was not flood but drought. In a narrative that symbolizes this climatic peril, the Yellow Emperor battled with his rival Chi You, a demon-like figure with four arms and the bronze head of a bull who commanded fire and drought. Chi You was said to have originally been the Yellow

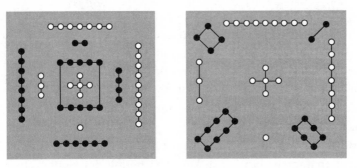

The *hetu* (left) and *luo shu* (right), the two symbolic schemes of Chinese divination.

Emperor's minister of waters, and he often featured in later drought-prevention rituals. So too did the goddess or drought demon named Ba, sometimes said to have been the Yellow Emperor's daughter, who at one time acted as Chi You's assistant but aided the Yellow Emperor in the fight against his rival. When Chi You called up a mighty storm during the epic Battle of Zhuolu, waged somewhere around the present-day border of Hebei and Liaoning on the North China Plain, Ba arrived to sweep the storm clouds away. Even then, it seemed, water was an element of warfare.

The myth of the Shang

Today these legends are often interpreted as a mythical retelling not of a battle against nature but of conflicts between rival tribes. The Yellow Emperor represents the ancient tribes of the Huaxia, thought to be the progenitors of the Han peoples, while Chi You leads the Jiuli ('Nine Li') tribes, sometimes identified with southern ethnic minority groups. This rationalization, aiming to marry legend to the 'official' archaeological history of Chinese civilization, was itself a kind of myth, since that history drew on the questionable authority of Sima Qian (*c.*145–86 BC), the first great historian of China, who served as astrologer in the court of the Han dynasty.

Sima Qian initiated the tradition of imperial dynastic succession that was so useful as an organizing scheme, indeed almost a moral rationale, for the unfolding of Chinese history. By situating the Han dynasty as the heir to previous 'unified' regimes, he was able to imply that there was something natural and even inevitable to its authority.

Sima Qian's history was also a narrative of cultural and ethnic dominance and prejudice. To him and to other historians before and after, the south of China – the middle Yangtze basin (the kingdom known as Chu), the delta region (Wu), Sichuan (Shu) and remote, jungle-strewn Yunnan – was still an uncivilized region where the populace was darker and barely human, their ways degraded and uncouth. 'Beware the songs of Chu,' warned Confucius, 'for they are licentious.' It was alluring, this moist and fertile land rich in fruit and fish, but that very abundance kept the people indolent. 'The people are lazy and poor and do not bother to accumulate wealth', said Sima Qian. Remnants of this north–south prejudice still linger.

Sima Qian's picture of historical succession was no less useful to the ideology of twentieth-century China than it was to the Han emperors. In the state-sanctioned modern history, Chinese culture began in the basin of the lower Yellow River, where a shamanic culture called the Shang dynasty – the dynastic label is barely more than a euphemism here – flourished around 1500–1100 BC.

By asserting that this Yellow River civilization gradually expanded over all of China, Chinese historians could assert a cultural unity that valorized assertions of nationhood in the face of ethnic tensions on the country's borders in Tibet, Yunnan and Xinjiang. In this way, national identity was yoked to the river and its landscape. When archaeological excavations at Zhoukoudian, south-west of Beijing, in the 1920s unearthed the remains of early human ancestors of the genus *Homo* around a quarter of a million years old, this so-called Peking Man served as an Adam for the modern creation myth of China: an alleged scientific verification of the idea of national ancestry from the Yellow Emperor.

But archaeology is too much a science to be dictated by political expediency. Over the past four or five decades this official story has been challenged, and it is now fairly comprehensively discounted by discoveries of rather advanced communities in ancient China that seem to have quite independent origins from the Yellow River culture. The traditional narrative became unsupportable after the discovery in 1986 of a sophisticated Bronze Age culture at Sanxingdui, near Chengdu in Sichuan. Contemporaneous with the so-called Shang dynasty of the twelfth and eleventh centuries BC, the Sanxingdui artefacts have a distinct artistic style that belies any connection of the two. Another culture different from that in the Yellow River basin seems to have evolved in the Yangtze valley.* A thousand flowers, it seems, really did bloom in the Chinese Stone and Bronze ages.

Some Chinese archaeologists have nevertheless tried to link archaeological finds to the conventional dynastic narrative, in which the Shang was preceded by the first dynasty to emerge from legend, the Xia (of which we will hear more shortly). In this reading, the Xia dynasty is ascribed to a Bronze Age culture of about 1900–1500 BC whose relics have been unearthed at Erlitou in Henan, while the Shang dynasty is associated with bronze artefacts dated at around 1500–1300 BC from Erligang, near Zhengzhou in Henan. Both designations, however, are unsupported by hard evidence and are viewed sceptically by archaeologists outside of China.

Perhaps the biggest affront to this traditional historiography is the claim (controversial in any event) that a separate early culture existed in the Tarim basin of Xinjiang in north-west China from around 1800 BC. This idea, promoted by American sinologist Victor Mair, is based on the discoveries of mummified bodies of distinctly non-Chinese – indeed, virtually Celtic – appearance, with blond/red hair and 'Europoid' body shapes. DNA testing confirms that the mummies seem to have had a European origin. The discovery has been seized on by Uyghur nationalists to assert their ancestral independence from China. With ethnic tensions in the region constantly at flashpoint, some Chinese academics are equally determined to discredit the idea. 'Within China', one historian has said,

> a small group of ethnic separatists have taken advantage of this opportunity to stir up trouble and are acting like buffoons. Some of them have even styled themselves the descendants of these ancient 'white people' with the aim of dividing the motherland.

Like just about anything else in modern China, archaeology is a political matter.

---

* There have been false trails too. The precedence of Peking Man as the ancestral Chinese seemed to be challenged by the discovery in the mid-1980s of hominid-like fossil remains at Longgupo cave in Wushan county, in the Three Gorges region of the Yangtze valley. 'Wushan Man' seems to be at least 2 million years old. But he has since been re-identified as not of the *Homo* genus at all, being instead a now-extinct form of ape – rather putting paid to the triumphant claims of some locals in the middle Yangtze valley that the entire Chinese race sprang from there.

Water dragons and river gods

Water has a central place in Chinese myth and legend, whether in folk belief or the official quasi-histories. Every river, lake, pool and spring in China seems to be associated with some tale or deity. River gods often seem to represent very ancient animistic beliefs; their clashes with the heroes of the dynastic myths can therefore be interpreted anthropologically as representing the displacement of local traditions by the hegemony of an external culture. A river god named Ju (or Chu) Ling presided over the Yellow River and could split mountains and ease the flow for humankind's benefit. He became a Daoist figure: one source from the Song dynasty says that Ju Ling 'chanced to obtain the Way [*dao*] of Divine Prime Cause, and he could create mountains and rivers and send forth rivers and water courses'. The evil Yellow River deity He Bo was deemed to be angry when it flooded, and livestock and perhaps even people were sacrificed to placate him.

Water myths and traditions have many local variants, especially among ethnic minority groups, for whom they often serve to establish and validate norms of mutual cooperation and group identity. Water reminds them of who they are and where their obligations and allegiances lie. The Dai people of southern Yunnan use water for blessings in ceremonies of birth, marriage and death. Their annual Water Splashing Festival commemorates the killing of a vicious demon who kidnapped seven sisters to be his brides. The defiant sisters beheaded him, but when his head fell to the ground it burst into flame, and each sister had to take a turn holding it while the others splashed on water to extinguish the fire before it could spread over all the earth. The Hani, Deang and Naxi people fetch water from rivers at the dawn of the New Year to bring good luck and prosperity; for much the same reason, brides among the Jino and Wa people draw up water from a well early in the morning after they have married. The Han myths of Herculean water mastery contrast starkly with the way water resources feature in minority myths, where they are locations that must be carefully tended and respected and which bind communities.

Dragons are associated with water all over South East Asia. They were linked to rivers, to rain and to rainbows, which were sometimes depicted with a monstrous dragon's head at each end. Spring pools are often called dragon pools, and many have dragon temples on their shores: such veneration no doubt ensures that the precious source of water is respected. Dragons, like rivers, were also symbols of fertility, both for crops and for humans. There are many folk tales of women who go down to the riverbank to wash clothes and come back inseminated by a dragon. The social convenience of such explanations in disguising sexual indiscretions is obvious enough, even if the women would allegedly then give birth to dragon children. Alternatively, they might cast themselves into the waters to drown their disgrace: an 'honourable' solution to the predicament that might leave the unfortunate victims venerated as water spirits.

Dragons made the great rivers of China; indeed, the rivers *are* dragons. Once, it is said, the seas were ruled by the Jade Emperor, Lord of Heaven, who appointed four dragons – Long, Yellow, Black and Pearl – to watch over the people and bring rain to water the crops. But the Jade Emperor's new wife enchanted him to pay attention only to her and to neglect the care of humankind. Despairing when rain failed to arrive, the people appealed to the dragons, who went to see the emperor in heaven. With some irritation, he sent them away with a vague promise that he would send rain soon. But he forgot, and still no water came from the sky. With no crops to harvest, the people ate roots, bark, even white clay.

Black Dragon Pool (Heilongtan) at Lijiang in Yunnan province. The temple was built in 1737 by the local Naxi people.

Dismayed by their hardship, the four dragons took pity. They bore seawater into the clouds, from where it fell as sweet rain. Seeing this, the Jade Emperor was furious that the dragons had made rain without his permission, and he commanded the mountain god to imprison each of them under a great mountain. Determined to continue to bring water to the people, the dragons transformed themselves into rivers that flowed down from the mountainside. Long and Yellow speak for themselves; Pearl flows through the southern provinces of Guangxi and Guangdong, emptying into the South China Sea, while Black is Heilongjiang (literally 'Black Dragon River') in the far north.

As this story suggests, the danger for early Chinese culture in the Yellow River valley was not so much flood as drought. At one time, it was said, ten suns floated free from the world tree Xi He, and they threatened to dry up the earth and turn the fields barren, scorching the sheaves of grain and making plants wilt. A skilled archer named Yi, sent to aid humankind by the God of the Eastern Heavens Di Jun, was commanded by the sage-king Yao (one of the Five Emperors) to shoot down the nine extraneous suns. The fireballs descended when struck by Yi's arrows and became a single flaming island far out to sea, named Wo Jiao, which evaporates all the seawater that touches it. This is why the seas, although constantly fed by the discharge of the mighty rivers, never overflow.

In some versions of this story, Yi overstepped the mark: the suns were actually Di Jun's sons, each of them a 'sun-bird', and the god had intended that Yi do no more than scare them out of their mischief. Yi was therefore banished from heaven, and subsequently becomes a Hercules figure who battled fearsome beasts and spirits on humankind's behalf: he is said to have shot and killed the Lord of the Yellow River.

Drought, like flood, was a sign of heavenly displeasure. After the Xia dynasty was overthrown by the first Shang ruler, Cheng Tang (traditionally in 1675 BC), the land was gripped by a drought that destroyed harvests for five years in a row. Poor families were forced to sell their children to survive – but the virtuous Emperor Tang minted and distributed gold coins so that the children could be bought back. Finally he went to a sacred mulberry forest to abase himself before heaven and to offer himself as a sacrifice. Then, as a Qin dynasty text called the *Spring and Autumn Annals of Master Lü* relates, 'the people rejoiced, for there was a heavy fall of rain'.

The tilting of the earth

According to anthropologist Bronislaw Malinowski, myth is 'a charter for social action': a cultural instruction for what needs to be done. Sigmund Freud's view that myths express unconscious fears and desires is not incompatible with that notion, and the flood myth of China fits both points of view rather well.

It's common knowledge that flood myths are universal – nearly every culture has one – but no one can tell you why. The location of many early civilizations next to great rivers that are prone to flooding suggests one obvious answer; it's tempting also to detect here a residual memory of the immense rise in global sea levels as the ice sheets thawed at the end of the last ice age around 11,000 years ago. But the image of the flood carries a powerful symbolic function. In contrast to the creation myths that begin before humans exist and which have a cosmogenic role of establishing a universe of gods and heavens, the flood wipes the slate clean for the emergence of human societies and institutions. It supplies a rationale for the here and now of the myth-makers themselves, a justification for the way things are arranged in their era – and a motivation for the establishment of a new covenant between heaven and earth, gods and humankind. It offers a Year Zero for the construction of genealogies, as the earth becomes repopulated and kinship and tribal structures begin to emerge.

In China, however, floods did not need to draw on either symbolism or ancient tribal memory. Floods on a mythic scale were still a part of everyday life, and remained a problem that had to be confronted and solved with every cycle of the seasons. This makes the flood myth of China unlike that of any other nation on earth. 'Probably no other people in the world', says Joseph Needham, 'have preserved a mass of legendary material into which it is so clearly possible to trace back the engineering problems of remote times.' The fact that these remain problems today sustains the cultural relevance of Chinese mythology in a way that is similarly unparalleled.

In many flood myths the uncontrolled waters have a moral agency. In the Noah legend, their message is clear: the waters descend as

God's punishment for humankind's wickedness, and only the virtuous man and his family escape this retribution. That narrative not only commends the listener to observe proper respect for God's authority, but also imputes a moral superiority, and therefore a right to rule, to the descendants of the survivors.

But in China the nature of water itself holds a particularly instructive message. That which causes the flood is violent, untamed, chaotic: it is 'bad' water. And it gets to be this way by a misdeed, a rebellious and criminal act. The stability of society depended on bringing this water under control: channelling and pacifying it, making it orderly and 'good'. Whoever could do this would be, virtually by definition, a virtuous person. In this way, water management becomes a moral issue: a successful system of control becomes not just a useful but an exemplary action, and the person responsible for this strategy demonstrates a right to rule. According to historian Mark Edward Lewis, 'every aspect of the Chinese flood myths . . . converged in meditations on the nature of the ruler and the justification of his authority'. A fear of disorder and chaos has always haunted China's rulers, but it's understandable that this metaphor should have been particularly stressed during the Eastern Zhou era when these myths were formalized. It was a highly turbulent period: from the fifth to the third century BC, seven warring states vied for control of the region loosely congruent with modern China. Not until the formation of the Qin dynasty in 221 BC did that struggle yield a unified state. The taming of the Great Flood represents a dream of social order in a time of turmoil.

The Great Flood took place during the reign of the Emperor Yao, whose era is traditionally assigned to 2350–2250 BC. In the *Shang shu* (*Classic of History*), one of the five canonical texts of ancient Chinese literature and dating from the Warring States period, Yao is described as a sun god and is credited with cosmogenic tasks: he orders time and space by situating four brothers at the cardinal points of the world to supervise the sun's movements.

In the *Shang shu* the Great Flood is presented as a neutral act of nature:

Like endless boiling water, the flood is pouring forth destruction. Boundless and overwhelming, it overtops hills and mountains. Rising and ever rising, it threatens the very heavens. How the people must be groaning and suffering!

But more commonly the flood is a consequence of transgression. According to one account, the deluge was triggered by rebellion of a king or demon named Gong Gong, who fought with one of the Five Emperors, Zhuan Xu (a grandson of the Yellow Emperor in Sima Qian's version), to gain a place in heaven. In that struggle the mountain Bu Zhou, one of the Pillars of Heaven that held the sky above the earth, was shattered. Through the hole created by this rupture, the waters of heaven came pouring, so that people had to flee their homes and climb trees or ascend hills to escape. Other calamities followed this breach in the Wall of Heaven: raging fires and ravenous beasts that wandered the land. The collapse made the earth tilt to the south-east, explaining why all China's rivers run in this direction. Even after the goddess Nü Gua intervened to repair the damage, the floodwaters covered the land. Some say Nü Gua used a mysterious 'five-coloured stone' for this feat, others that she plugged the gap with her own body. She is said to have piled up reeds, wood and stones carried from the western mountains to make dykes against the waters; one legend says she did this in the form of a reincarnated bird (*jingwei*) after drowning in the Eastern Sea.

You don't get far into the Chinese flood myth before the discrepancies among the various versions begin to complicate the picture. Gong Gong is sometimes presented as a sea monster or a water demon with a reputation as a flood-monger. The Han-era text known as the *Guanzi** doesn't blame him for the flood itself, but says only that he was a king who tried to exploit the disaster by seizing control of the world. This lack of differentiation between deities, demons, beasts and government officials speaks of a complex interplay between folk legend, politicized history-making and the apotheosis of authority figures.

In the same way, we can find these myths being adapted to offer naturalistic explicatory schemes while simultaneously being tailored to validate power structures. The links between riparian engineering

---

* This text is named after Guan Zhong, the prime minister of the legendary Duke Huan of Qi from the seventh century BC, but it was collected by the Han scholar Liu Xiang, mostly from fourth-century-BC sources.

strategy and good morality are clear in a version of the flood myth which states that:

> Kung Kung [Gong Gong] abandoned the Way [*dao*] . . . He wanted to dam the hundred rivers, reduce the highest ground, and block up the low-lying ground, and so he damaged the world. But august Heaven opposed his good fortune and the common people refused to help him. Disaster and disorder sprang up everywhere and Kung Kung was destroyed.

Because Gong Gong's name translates as 'communal labour', there have been suggestions that he represents the defeat of a collectivist society by a hierarchical one. In any event, be he water demon, rebel leader, hubristic usurper or moral outcast, he was eventually eliminated: either destroyed or humiliated. Yet his legacy remained. China was covered with water, disrupting social order until the flood could be overcome. That was accomplished by the first and greatest of China's water heroes, Da Yü or 'the Great Yü'.

The Tracks of Yü

In the *Shang shu*, Gong Gong is not (literally) demonized, but is described merely as the first person Yao considers for the role of flood-control engineer. He is swiftly rejected after his strategies are deemed unsuitable. The next candidate is a man called Gun, who attempts to quell the waters for nine years until it is clear that he is getting nowhere. Gun's water-management philosophy is flawed too: he tries to block the waters rather than helping them to flow and drain. The *Shan hai jing* (*Classic of Mountains and Seas*) makes Gun's error a crime: he steals the supreme deity's 'self-renewing soil' (*xi rang*), a mysterious substance that could apparently swell to dam floodwaters. This theft prompts the god to command his death.

At some stage – sources are divided on the exact timing – Yao appointed his own successor, a faithful and able minister named Shun who is another of the five sage-kings. This Shun is an interesting character, seeming quite distinct from the other semi-divine sages. Some scholars think he was derived from a water spirit associated with creatures of the marshes. He was apparently close to nature in any event: according to the Confucian text called the *Mencius*, he was at first 'scarcely different from a wild man of the mountains'. Shun exemplifies the blurred distinction between human and animal that is so important for early Chinese moral philosophy: this ambiguity pertained in a pre-civilized state of affairs said to exist before the flood, establishing the notion that taming the waters was the foundational step for civilization itself. Shun was responsible for taming the birds and beasts. His mother, it is alleged, was impregnated by a rainbow, and he himself had the face of a dragon, referring to their roles as water symbols.* The *Mencius* seems almost to make Shun an embodiment of the flood itself:

> When he heard a single good word or witnessed a single good deed, then it was as if one had opened up the Yangtze or the Yellow River. As he poured out with great force, nothing could check him.

Shun was initially appointed by Yao as a co-ruler to share the burden of dealing with the flood. He did not seem like promising material – a common man (although it was later felt necessary to trace his lineage to the Yellow Emperor), his father was stupid, his stepmother arrogant, and his half-brother conceited. Yet Shun's filial conduct was so exemplary that this difficult family was able to live in harmony, and he performed his administrative duties with great skill. His suitability for this demanding official role was put to the test. Yao granted Shun both his daughters in marriage, reasoning that if you can't manage a household then how can you manage a flood? Shun coped with that just fine. He was also exposed to an ordeal in the wilderness in which he had to endure 'violent winds and terrible thunderstorms'. Proving equal to these tasks, he was invited to share Yao's throne. The older emperor became little more than a figurehead; it was now Shun who held the reins of the state.

He considered what might be learnt from Gun's abject failure. Gun had wanted to build great dykes to contain the waters, and

---

* In Shun we see most explicitly the common belief that all the sage-kings, including Yao and Yü, were preternatural demigods, the offspring of a mortal woman and a god or spirit. Some think that Yü was originally a fish spirit (his name is a homophone of the Chinese word for fish); there is an ironic echo of that idea in the popular Chinese saying that 'If it hadn't been for Yü, we would be fish.'

when they overflowed Gun insisted that they simply needed to be raised higher. Shun saw that this wasn't going to work, not least because it was all but impossible to coordinate the civilian labour it required. So the first act of taming the flood was to institute a new kind of civic society, organized along well-defined principles. Shun standardized the calendar, the system of weights and measures, and the written language. He introduced ceremonies for death, marriage, ancestor worship and warfare. He divided up the flooded land into twelve administrative regions. Above all he sought to establish a stable, harmonious and well-governed society: a template that every succeeding dynasty sought to emulate. According to Wu Kuo-Chen*:

> Shun gave the Chinese a heritage perhaps even more enduring than what he himself had at first thought possible. From his time on down to this day, these ideals – harmony and propriety [li] – have been deeply embedded in the minds of the Chinese people.

First Shun had to get rid of Gun, who responded indignantly when confronted with his shortcomings. On what grounds, he demanded to know, did Shun, a mere commoner with no experience of statecraft, justify his authority? Shun's answer was one of Maoist bluntness: the argumentative engineer was banished in short order to the remote Feather Mountain on the south-eastern coast.

With Gun removed, his son Yü stepped up to take his place. Yü is not 'China's Noah'. He is not favoured by the gods or by fate, he is not a lone survivor and the father of all who came after, and most importantly, he doesn't passively ride out the flood. He takes matters into his own hands: it is no exaggeration to call him a hydraulic engineer. He is generally assumed to correspond with some real ancient ruler, but he bridges the age of gods and the age of men, personifying the moment when legend starts to become history. He is universally venerated in China as one of the Five Emperors, the founder of the Xia dynasty that is deemed to commence around 2200–2100 BC. It was

---

* As well as being a historian, Wu was also active in Chinese politics, and was himself familiar with water management. As mayor of Hankou (a part of what is today Wuhan) in 1936 he oversaw the raising of the Yangtze dyke to avert a threatened flood.

his domination of the floodwaters that won him this absolute respect and authority, and China's leaders have never forgotten that.

Yü laboured for thirteen years, exemplifying the selfless devotion that the Chinese state has traditionally expected of its servants. As described in the document from the Warring States period called the *Yü gong* (*Tribute of Yü*), he began at a hill called Hukou on the Yellow River (in modern Shanxi), where the seething waters were threatening to flood Yao's capital. Systematically, Yü dug and deepened channels on the Yellow and then the three other great rivers, the Ji,* Huai and Yangtze. Ceaselessly he paced back and forth over the land, supervising the carving of passages through mountains and the dredging and redirecting of rivers. He expelled wild nature, its snakes and its dragons, back to the swamps. The hairs on his arms and legs were burnt away, and he walked so far that he became lame, proceeding in an awkward manner known as the Gait of Yü – a posture said to be adopted by the shamans of the Shang dynasty. His journeying made him a patron of travellers. Several times Yü passed close enough to his home to hear his poor wife (whom he married just four days before starting his work) and children crying for his return, but he ignored their tears and continued on his way.

Why did Yü succeed where others failed? This is the issue that unites the apparently disparate notions of moral conduct, water management and state authority; in a sense, nearly everything else connected to water in Chinese culture follows from it. According to the *Mencius*, Yü enabled the floodwaters to find their natural way, their *dao*. This was not, however, by any means a matter of laissez-faire. It involved heroic engineering, in the description of which the *Mencius* becomes almost a technical report. The Nine Rivers were dredged, the channels of the Ru and Han rivers were opened up and those of the Huai and Si were 'cleared out'.

What Yü did wasn't merely efficacious but virtuous. It was the natural way. His predecessors, in contrast, failed precisely because their strategies violated the natural order. Gong Gong had tried to block rivers and topple mountains, to obstruct rather than open up.†

---

* This river is mentioned by Sima Qian, who identifies it as the ancestral home of the Yellow Emperor's tribe in north-east China. But it is not clear exactly where or what it was; some link it with the Fen River in Shanxi.

† In accounts that draw this contrast, Gong Gong takes the role of Yü's inadequate father Gun, rather than being the instigator of the whole mess. This is a hallmark

The French sinologist Marcel Granet suggested that the difference between the approaches of Gong Gong and/or Gun and that of Yü – whether to build the dykes high or to deepen the river channels – represents a conflict between rival schools of thought in ancient China about hydraulic engineering of much the same nature as that later identified between 'Confucian' and 'Daoist' engineers (see page 163). But the message goes beyond hydraulic engineering. It is a warning against any 'unnatural' moral conduct that might prevent one's own dynamic energy (*qi*) from flowing freely. The way to manage water is the way to manage ourselves.

Dividing the land, making a dynasty

By draining away the floodwaters, Yü revealed the land afresh: now not a part of nature but the product of human labour. It had become, in short, a state – and as such, it needed to be organized. By supervising this work, Yü established the basis for a bureaucratic system of governance. As Shun's 'minister of works' he measured up the lands and divided the country into Nine Provinces. He shored up the so-called Nine Marshes, commanded the cultivation (meaning the deforestation) of the Nine Mountains, and opened up the concourses of the Four Seas. He was, in effect, implementing an agricultural and administrative policy.

Other officials were (the legend has it) appointed to set up social institutions and laws: to create an orderly state from what was previously inundated chaos. They formalized social relations, taught agriculture, arts and crafts, and drew up rituals and customs. Yü's reforms included the introduction of surnames as a way of specifying familial relationships. He implemented a system of tax levies, calculated according to a sophisticated calculus of soil type, population density and irrigation measures. He oversaw the creation of tribute payments to the emperor, paid in furs, silk, pearls, ivory, jade and other luxuries. In this way, the legend of Yü serves also as a validating myth for many essential features of Chinese agrarian society: the regulation of waterways, the distribution of land, the hierarchies of command and

---

of the evolutionary and patchwork nature of the myths: it may have been only at a late stage that Gun was inserted and Yü made his son.

obligation, and the use of organized labour for state projects. It is in the *Yü gong* that the first reference appears to this kingdom as Zhongguo: the Middle Kingdom. In other words, Yü's labours can be considered to constitute the foundational myth of the Chinese state. As Mark Lewis puts it, 'The flood myths that developed around the figures of Yü and Nü Gua provide a comprehensive mythology for the origins of virtually all the units in terms of which the early Chinese articulated their notions of a structured space.'

Yü was entrusted by heaven with the principles by which the Chinese nation should be ruled.* Chief among these was the Great Plan (*Hong fan*), a set of Nine Categories that established how the 'various virtues' of humankind should be regulated by a sovereign. You'll notice the recurrence of this auspicious number nine. Few habits are more distinctively Chinese than this insistent enumeration of every cultural entity, facet or trait. And indeed, to venture into the definitions of these Categories is truly to enter the thickets of this penchant for list-making – they range from the eight proper governmental offices to the 'five arrangements of time'. They are perhaps best regarded as an illustration, first, of the importance of agreed principles in early Chinese statecraft, and second, of how little distinction there was in Chinese thought between metaphysics, systems of governance, and personal morality. In pleasing the Lord of Heaven by his proper management of the flood, Yü demonstrates the supremely Confucian notion of right conduct. As the ruler of the ancient state of Wu was told by one of his noble advisers, heaven withheld the Great Plan from Gun because his poor hydraulic engineering caused chaos among the elements, but Yü was favoured and became the quintessential emperor: the first ruler of the Xia dynasty. Throughout the Shang, Zhou and Warring States periods, the Chinese people thought of themselves as the Xia people: the descendants of Yü's dynasty. It was not until the Han dynasty that the majority race of China adopted this designation instead.

The Yü myth also advises on the principles of succession. The ineffectual Gun is followed by his irreproachable son Yü – indeed, some say that Gun was so wicked and dissipated that he was slain by

---

* In some accounts, Yü also received the divinatory *luo shu* magic square from the Yellow River, just as Fu Xi received the *hetu* on which the *I Ching* was based.

the gods, whereupon Yü emerged from his body. And because the emperor Shun doubts the virtue of his own son Shang Jun, he passes on the throne instead to the meritorious Yü – just as Yao selects Shun to be his heir instead of *his* son, the degenerate Dan Zhu, who is described as 'abusive and disputatious'. The 'bad son' was commonly exiled to the furthest fringes of the kingdom – Dan Zhu was sent away to the region of the Dan River in the south – where he might be transformed into a threatening demon to ward off invaders. This inversion of morality from one generation to the next – an excellent ruler produces corrupt offspring and vice versa – served to discredit any notion of hereditary rule in favour of meritocratic succession by worthy officials. That was the principle later recommended by Confucius for state appointments, although he prudently stopped short of advocating it for emperors.

Yet by founding a dynasty and transferring kingship to his own son Qi, didn't Yü undermine this principle of succession? Well, yes – and it wasn't necessarily a good decision. Some say that Qi's son Wu Guan was a wicked man, others that Qi himself was deviant. The stories reflect a real unease about the best way to hand on power, which even the end of imperial China in the modern era has done nothing to dispel. But this wasn't just about kingship: the father/son themes of the Chinese flood myth are also exploring ideas about *kinship*, with the macrocosm of the kingdom posing the same questions about morality, lineage and filial piety as did the microcosm of the household.

The uses of myth

Following the disintegration of the Zhou dynasty in the fifth century BC, the hydraulic engineers of the Warring States period claimed that they were 'carrying on the work of Yü'. An association with the virtuous Yü was useful for any ruler. It is not hard to see why the first Qin Emperor should have approved a version of this myth in which his ancestor Da Fei was Yü's erstwhile assistant. When King Ling of Zhou in the sixth century BC wished to block up a river that threatened to flood his palace, his heir apparent warns him that this was not the way of Yü:

This is not permissible. I have heard that those in antiquity who nourished their people did not topple mountains, raise up lowland wastes,
block rivers, nor drain swamps. Mountains are the gathering of earth.
Lowland wastes are where creatures take shelter. Rivers are where
energy [*qi*] is guided. Swamps are the amassing of water . . . The
ancient sages paid attention entirely to this.

In other words, nature has ways that must be respected. Just as good
water management means discerning and respecting the *dao* of the
rivers, so a wise ruler will identify and observe the natural tendencies
of his people. Then all will be peaceful and orderly.

Da Yü became the legitimizing figure of water management – and
by extension, of all state authority. When the Ming Yongle Emperor
in the early fifteenth century was anxious to establish his right to rule
after he had overthrown his nephew the Jianwen Emperor, it wasn't
enough to rewrite the official records. His favourite minister, Xia Yuanji,
advocated new hydraulic practices, such as dredging, on the grounds
that these followed the principles of the 'Three Rivers' allegedly
described in the *Tribute of Yü*. Xia's plans stated explicitly how these
'old' techniques situated the Yongle Emperor as a sage-king in the
mould of Da Yü:

Thinking retrospectively of the Three Rivers, they were dug by the
Holy Yü. They have been neglected for an extensive period of time,
and are now about to dry out. Thus paddy fields are damaged and the
spirit of the River Deity is withering. Fortunately, now the sage has
re-emerged in the world. Day and night he is worried about, and
dedicating himself to, [the people's welfare]. He ordered me to drain
and guide [the water].

In truth, it wasn't even clear what the legendary Three Rivers were, let
alone quite what Yü's strategy to fix them had been. Nor was it clear
that the lower Yangtze delta, to where on this pretext Xia was sent to
put things in order, really needed fixing at all. But what was required
at that point in Yongle's reign was a gesture that connected him to the
ancient water god. Xia's methods were quietly abandoned later when
there was no longer a need for this mythological grand narrative. Instead

The Great Yü presides over the Sanmenxia Dam on the Yellow River at the border of Henan and Shanxi.

of labour-intensive engineering of the rivers to prevent flooding – which risked the unpopularity of conscripted work and was likely to be of limited effect anyway – the victims of such disasters were simply given aid and tax relief in a show of benevolence.

The historian David Pietz says that 'the sanctioning power of myths, adapted and retold to legitimize political authority, was expressed in a host of water-management projects throughout history'. This political manipulation of legend is as active today as ever it was – note, for instance, how Da Yü's authority was mobilized to validate the Sanmenxia Dam in the late 1950s (see page 23). More recently, a garden has been created on the cliff face overlooking the dam, dedicated to Yü and depicting him as a gigantic Hercules-like figure standing watch over China's most notorious torrent.

The glorification of Yü is part of a more general political rehabilitation of China's traditional past. When the Qingming Festival, the 'Grave-sweeping Day' on which ancestors are honoured, was officially reinstated in 2006 after being banned during the Mao era, the celebrations in Shaoxing, said to be where Yü died and was buried, were reported to include 'ancient rituals' to venerate the water hero. And when President Jiang Zemin, an engineer, inscribed the characters on the gateway of the Da Yü Mausoleum in Shaoxing in 1995, he knew what a deep well of trust and veneration he was drawing from. The press duly anointed him as 'the new Da Yü': a leader who derived his authority from privileged knowledge about controlling the waters.

3 Finding the Way

Water as Source and Metaphor in Daoism and Confucianism

> As regards the people who protect and manage the dykes and channels of the nine rivers and the four lakes, they are the same in all ages; they did not learn their business from Yü the Great, they learnt it from the waters.
>
> *Shenzi, c.* fourth century BC

'Water! Oh, water!' Confucius was said to have been fond of exclaiming. But what did he mean by it, his follower Xuzi wondered? What was so admirable about water?

The fourth-century-BC scholar Mengzi (whose name is usually romanized to Mencius), the most celebrated of the great master's interpreters, tried to explain. Water, he wrote,

> from an ample source comes tumbling down, day and night without ceasing, going forward only after all the hollows are filled, and then draining into the sea. Anything that has an ample source is like this. What Confucius saw in water is just this and nothing more. If a thing has no source, it is like rainwater that collects after a downpour in the seventh and eighth months. It may fill all the gullies, but we can stand and wait for it to dry up. Thus a gentleman is ashamed of an exaggerated reputation.

This doesn't seem to help us very much, does it? How exactly do rain-filled gullies show us what a gentleman should attend to in his reputation?

That's so often the way with classical Chinese thought, whether it is Confucian, Daoist, or derived from one of the other philosophical traditions. The allusions and connections, the currents of logic, are subtle and sometimes unpredictable, and they are apt to leave us as baffled as poor Xuzi. Yet they read like poetry.

Put aside for a moment what it is that Mencius wants to convey about the quality of a gentleman, and consider simply that *water* supplies him with the imagery for exploring it. Is that surprising? Surely, you might say, water is so protean a medium that it can be made to fit any purpose. It can be wild and raging or tranquil and limpid. It can collect and stagnate, or it can flow with abandon. It reflects like a mirror, it is transparent, or – like the Yellow River – it can be murky and opaque. Water may be both transient and continuous. You can make of it whatever you will.

But this is precisely the point. It is both because of its changeable nature and because that nature was so ubiquitous in daily life that water was *good for thinking with*. That situation in itself was not unique to China. All cultures tend to incorporate their material world into the metaphors they use for living, and water is a particularly rich and versatile medium for doing so. The French philosopher Gaston Bachelard has identified many of its mythic associations: it cleanses, purifies, nurtures and destroys. Narcissus loses himself in water's placid mirror. It is a deathly embrace: 'Too much of water hast thou, poor Ophelia', Laertes laments. There is an energetic struggle with water that Bachelard christens the Swinburne complex, after that poet's obsession with immersion and swimming (a masochistic impulse, Bachelard claims). Water has a hygienic morality and a fathomless Gothic horror, plumbed by Poe. Nothing in all this seems to distinguish Chinese thought as unusual in its readiness to reach for water images.

But you will struggle to find another place in the world where water is as *indispensable* to thinking as it is in China, and particularly in the works of its great philosophers and poets. Mencius's comment, however opaque it might otherwise seem, implies that there is nothing superficial or idly mimetic about the roles water serves here: it appears to embody some deep principles that guide the Confucian mode of thought. This is equally true, if not still more so, of the other central system of philosophy in Chinese tradition,

Daoism.* In both cases, water isn't simply a handy source of comparison and illustration: it is itself a virtuous substance, bearing a force of moral instruction. 'Water is the highest good', wrote Lao Tzu in the *Dao de jing*, the foundational work of Daoism.

For this reason, calling China the Water Kingdom means something far deeper than that water has dictated the patterns and customs of domestic, civic and political life. Or rather, we might say that water had this social significance for precisely the same reasons as it was philosophically fundamental: there was no escaping its importance. 'As the years went by', Joseph Needham writes, 'the whole of Chinese theoretical thought came to be permeated by certain ideas proper to the control of water-courses, so important a feature of the civilization.'

This is the truly extraordinary thing. Not only did daily experience of water provide metaphors for philosophical thought, but the philosophy in turn influenced practical affairs: there evolved an intimate connection between hydraulic engineering, governance, moral rectitude and metaphysical speculation that has no parallel anywhere in the world.

Writing with water

The Chinese character for water, *shui* (水), is one of the most beautiful. It has the 'almost-symmetry' to which our aesthetic sense seems especially attuned. Its little basal hook suggests dynamism, while the splayed legs give a reassuring stability. No doubt part of this character's appeal for students of Chinese is also that, like many of the most common *hanzi*, it is relatively simple, accomplished with just four strokes.

Shui is one of the five classical Chinese elements (*wu xing*), along with wood, fire, earth and metal. As such, it represents a principle or ideal as much as a physical substance: in *shui yin* (水银, mercury) it denotes fluidity, in *shuijing* (水晶, crystal) it is the transparency and clarity of water that is invoked. In classical texts *shui* may refer to a phenomenon such as a river or flood (*da shui*, 'big water'), much as the biblical use of 'water' denotes a generic, primordial agent that is not necessarily the same as the substance we drink and in which we bathe: 'And the Spirit of God moved upon the face of the waters.'

---

* There is more to be said in this context about Buddhism too, which has had a profound influence in China. But that was, of course, ultimately imported from outside.

But *shui* does not, for the most part, allude to some vast undifferentiated mass of oceanic proportions. For most Chinese people even today, the water that they encounter is not the sea but that in great rivers, in irrigation ditches, in pools of rain and mists on the mountaintops. The character itself testifies to this: it appears to be a stylized central stream confined by channels. This is pictorially explicit in the early forms of the character from the Shang dynasty (*c.*1700–1100 BC):

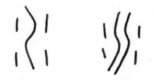

In the symbols for natural springs (*quan*) the flow emerges from a source:

This structure is retained in the modern character, in which the character *shui* flows down from the *bai* (white, 白) radical: 泉.

These early representational forms of the Chinese script appear on so-called 'oracle bones' found in the region of Anyang in Henan province, which were created by Bronze Age settlers in the Yellow River valley sometime around the fifteenth to the eleventh century BC. As we saw earlier, their conventional attribution to a culture labelled the Shang dynasty is designed primarily to fit comfortably within the Chinese historical tradition of dynastic succession, implying a kingdom and social structure rather more extensive and grand than was probably the case. The life of these Shang people was precarious. They appear to have been an agrarian culture, yet irrigation was all but unknown to them. The society extended no further than northern Henan and south-east Shanxi, and was likely to have been just one Bronze Age culture among many in northern China. It is the very existence of the script on the oracle bones, with its clear connections to some of the later written characters, that allows this society to serve a notional ancestral role for Chinese civilization.

The oracle bones had been known for centuries in the region, where they had been sold as 'dragon bones' for making medicines. It wasn't until the start of the twentieth century that archaeologists realized their significance, and excavations at Anyang unearthed thousands of them. They are ritual artefacts: animal bones and tortoise shells used for divination and augury. They were heated until they cracked; in the patterns of the cracking, shamans would read oracular messages. To our good fortune, these interpretations were engraved into the bones themselves in the elegant, pictographic Shang script. Around 4,000 of these characters have been identified, but only half have been assigned (with varying degrees of confidence) a meaning.

The oracle-bone characters are often delightful in their stripped-down visual logic. A particular river such as the Huan would be denoted by a symbol with that phonetic sound (*huan*,) combined with the water element:

Often water is depicted as a series of short strokes, much as rain is conventionally shown in cartoons and children's pictures today. Thus 'bathing' shows a man () splashing in a vessel ():

while 'urination' is an exercise in shorthand of Miro-like invention:

These splashes are retained today in the 'water radical', which is one of the fragments from which *hanzi* characters are formed (see 'What's in a character?', page 73). The water radical (or) splashes out from the left of characters such as *he* (河, river), *jiang* (江, river), *hai* (海, sea), *tai* (汰, tide), *gou* (沟, ditch, channel or ravine), and, tellingly, *min* (泯, obliterate).* The stream or river radical (巛) likewise has a pictographic quality that speaks for itself.

* Here, as elsewhere, when writing in Pinyin I am not indicating the character's tone – that is, the way it is enunciated, which provides a crucial clue to the meaning.

What's in a character?

If the water character 水 is a pictogram, it is unusual in that regard. The number of Chinese characters that can be classified as pictographic is small: sun (*ri*: 日), moon (*yue*: 月), river plain (*chuan*: 川), mountain (*shan*: 山), for example. Most characters are now regarded as logographic: their shapes aren't stylized attempts to *draw* the word represented, but are merely a notation for denoting it.

All the same, while they might seem like random marks on first encounter, Chinese characters have a degree of organization and logic that offers clues to both meaning and vocalization. The most fundamental units of the writing system are individual strokes, each of which is made in a single gesture without lifting brush or pen from paper. From these atomistic ingredients are constructed certain radicals or *bushou*, each of which conveys a particular meaning. Some of the simpler characters, such as *ren* (人, person) and *dao* (刀, knife), consist of just a single radical. More complex *hanzi* may contain several radicals combined, generally either side by side or one atop the other. In side-by-side characters, the left-hand radical hints at meaning while the right-hand radical generally supplies a clue to the way the character sounds: it is a phonetic, not a semantic, label. As one gets to recognize these radicals, the characters reveal their inner form and composition, and become a little easier to interpret and remember.

Radicals are typically used in words that have some obvious link to their meaning. Some of the radicals are abbreviations of the corresponding character, although this relationship may be rather cryptic. Thus, for example, the person radical (亻), derived from *ren*, person (人), appears in the modifier for groups of people, *men* (们), the article particular to people, *wei* (位), and the pronoun for 'you', *ni* (你). The hand radical (扌), from *shou*, hand (手), appears in *da* (打, hit), *la* (拉, pull) and *tuo* (拖, drag). The logical structure of the written Chinese script offers a shred of hope for those trying to learn it, although the logic has its limits: it is far from obvious, for example, why a word such as 淫 (*yin*: lewd) would contain the water radical.

All the same, it can be telling where the water radical makes appearances. Most strikingly, the Chinese character for political power and governance, *zhi* (治), is comprised of fragments denoting water and a platform or stage (*tai*, 台), implying that state rule is a platform built on water: a concise statement of the key contention of this book. It is probably no coincidence either that the water radical features in the characters for law (*fa*, 法, originally having the broader meaning of 'standards') and benefaction (*ze*, 泽). The common negation word *mei* (没) can also serve as an adjective meaning 'drowned'.

More sophisticated concepts are often conveyed by a suitable juxtaposition of two or more *hanzi* into a composite word. This is in many cases entirely analogous to similar marriages of Latin or Greek roots in English, so that while for example *telephone* is a composite ('long distance voice') the meaning of which even Plato could have inferred, so too the Chinese equivalent *dianhua* (电话, 'electric speech') embraces concepts that would have been familiar even to Song dynasty scholars for whom *dian* meant 'lightning'. In this way, *shui* does sterling work in diverse concepts that contain something of the fluid: fruit (*shui guo*) and tears (*lei shui*: here *lei*, 泪, shows droplets spilling from the eye: *mu*, 目). Elsewhere the appearance of *shui* betrays the origin of a more fundamental concept, revealing how water made it concrete: *shuiping* (水平) connotes a level or standard.

Life spirit

A predilection for dualism isn't unique to China. The feminine *yin* (阴) and masculine *yang* (阳), and their many correlates, have ample analogues elsewhere. The utility of such schemes is obvious: polarities provide a scale on which all manner of phenomena can be ranked and organized. A balance of these complementary cosmic forces – of cold and hot, female and male, moon and sun – promotes universal harmony. As *yin* and *yang* 'mingle and join', says the third-century-BC text the *Zhuangzi*, 'from their conjunction comes to birth everything that lives'. The origin of the ancient *yinyang* symbol itself is not known,

Yin and *yang*: a fluid dynamism.

so it isn't possible to be sure what it depicts – but the fluidity, the mingling of eddies in flowing water is plain enough. One interpretation of the *Zhuangzi* identifies *yin* and *yang* with the states of water: '*Yin* in its highest form is freezing while *yang* in its highest form is boiling.' *Yin* embodies the ancient, cross-cultural association of water with the feminine, contrasting with the fiery masculinity of *yang*. After a severe flood in AD 813 the Tang emperor Xianzong decided that the cause must be an excess of feminine *yin*, and he expelled so many women from his palace that 200 wagons were needed to bear them all away.

There is a particularly beautiful aspect of the *yin/yang* dialectic that runs through all of Chinese thought and artistic expression, namely the contrast of mountains and water: *shan* and *shui*. The (male) mountains are permanent, symbolizing space; (female) water is changeable, a symbol of time. Mountains rise, waters descend. But they are symbiotic: rivers begin in mountains, and they are the sculptors of mountains. In this way, *shan* and *shui* can represent the entire cosmos. They were both sacred places and sites of ritual offerings, including (in the most ancient times) human sacrifices. They may also signify human characteristics: 'The intelligent find joy in water,' says Confucius in the *Analects*, 'the humane find joy in mountains. The intelligent are lively; the humane, still. The intelligent are happy; the humane, long-lived.' You can argue about which is better; but if you throw in your lot with water, you need to be smart.

Among the many other polarities represented by these two principles, the Han scholar Xu Shen's dictionary *Shuowen jiezi (Explaining Graphs and Analysing Characters, c.*AD 100) lists another that is dictated by China's distinctive topography: the south and north banks of a river. In most areas of the world these would seem somewhat arbitrary associations, but for the west-to-east-running rivers of China the south-facing banks are sunlit while those facing north are shaded – a distinction of great practical importance for peasant farmers in the river valleys.

The *yin/yang* dualism seems to have been consolidated around the third century BC, before which there was a more conventional opposition of fire and water. These two are kindred elements, of course: both flow, both may sustain life but may also destroy it. They are mutually destructive. The two characters, *shui* (水) and *huo* (火), are visually similar; in combination, they may represent calamity.

Both fire and water move of their own accord, because they are imbued with the vital quality of *qi* (气). A concept of this nature features in many ancient cosmologies, from the Hindu *prana* to the Western humours, *élan vital* or *spiritus mundi*. Yet *qi* again defies easy translation. The sinologist Angus Graham suggests that 'we could think of it in Western terms as pure energy', but with all due respect to Graham's immense learning this invites misinterpretation: it is a pre-scientific concept, and, like many such, therefore a notion whose primary value is intuitive. It's always wise to stay mindful of the words of the sinologist Charles Alexander Moore: 'Some Chinese terms are so complicated in meaning that there are no English equivalents for them and they therefore have to be transliterated.' This warning is especially pertinent in discussing Chinese philosophy.

What perhaps most distinguishes *qi* from cognate ideas in other cultures is its dependence on flow. In the system of Chinese medicine, the unobstructed flow of *qi* is essential for health, and by the same token a healthy watercourse is one in which there are no blockages. 'When fluent *qi* takes form, order is generated therein', says the *Xunzi*. There is in *qi* something of the same polyvalent sense as in the Greek notion of *pneuma*, which may be understood simultaneously as air, wind, breath or spirit. Today *qi* can be translated as air, gas or breath; weather (*tianqi*, 天气) is literally heaven's breath.

But it is really water, not air, that provides the foundational concept of *qi*. Water has the fluidity of *qi*; indeed, according to the fourth-century-BC text the *Guanzi*: 'Water is the blood and *qi* of the earth; it resembles what courses through the veins and arteries. So, it is said that water is the potential for everything.' The *Shuowen jiezi* dictionary defines *qi* as 'cloud vapour', and the early (Shang dynasty) forms of the character reflect this in the way they hover like stratus formations:

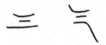

The traditional form of the modern character (氣) is also associated with clouds, water vapour and mist: it is really 'steam', rising from the cooking pot full of rice (*mi*, 米).* That is the beauty of this extraordinary language, in which the characters may occasionally blend the metaphysical with the practical not just conceptually but visually, and thereby connect profound and abstract concepts with the daily routine. The natural metaphors beneath the philosophical terminology can never be overlooked because the very act of writing displays and reinforces them.

Learning from nature

By the standards of most world religions, China has always been a godless nation. To the extent that early Chinese thought can be considered religious at all, it was based on what might be deemed a spiritualist or animistic tradition. One was obliged to pay respects to one's ancestor spirits, but nature too was alive with spirits: not exactly transcendental beings, but a part of the natural order.

While there was a supreme deity of sorts, his role was more that of an administrative figurehead – an overseer of creation, one might

---

* The older 'traditional' forms of many characters are more complex than the 'simplified' forms introduced in the 1950s and 60s in a drive to boost literacy. Traditional characters still predominate in Hong Kong and Taiwan. In the simplified modern character for 'steam' (also *qi*), it is distinguished from the more generic *qi* of air, gas or spirit by the simple measure of preceding it with the water radical (汽).

say – than a moral guardian. That is reflected in the use of the word *di* (帝) to denote him: it was the title of the dynastic emperors too. In the Shang period, a 'Lord on High' (Shangdi, 上 帝) was considered to control the rain and clouds. But throughout the Zhou dynasty Shangdi was generally equated simply with the *tian* (heavens), a supreme, impersonal organizing force. The Jade Emperor, who we encountered earlier, was, crudely speaking, the Daoist equivalent of Shangdi – or identified as his assistant and successor Yu Huang Shangdi – and was widely considered one of the three pure emanations of the *dao*, the ultimate source of all that exists.

These godlike beings did as emperors did (or should): they battled evil external forces (monsters and demons), looked after their people, and observed the *dao*. They didn't prescribe a moral order; neither did they imbue the cosmos with meaning. Yet people cannot do without those two things. If they can't be assigned to a personified intelligence, their roots must be sought elsewhere. The effective absence of a supreme moral authority allowed, if not compelled, an alternative cosmic teleology to be found. The sinologist Sarah Allan explains that, with no canon of religious precepts, Chinese philosophy took what she calls its 'root metaphors' from the natural world. The *I Ching* in particular exemplifies the Chinese manner of using naturalistic images to illustrate aspects of human thinking and conduct. By studying nature, one could understand the forces that shape human society. And of all the natural systems and structures that could offer instruction for human conduct, water was the most valuable.

Nowhere is that more true than in an elucidation of the concept of the *dao* (道), the 'Way' that is central to both Daoism and Confucianism. While Western scholars have often translated this word as a 'road' or 'path', Allan argues that in fact water provides its root metaphor. The Way is explicitly associated by Lao Tzu with a water-course or irrigation channel, which is sometimes denoted by the character *zhu* (注). (Today this character generally means 'to annotate or take note of', but its water radical betrays its root as a word meaning 'to pour or fill'.) The verb form of *dao* can refer explicitly to the act of cutting a channel so that the water that flows through it may attain its own *dao*, bringing benefits to all of nature.

The *dao* is one of those philosophical concepts that resists translation. It connotes a kind of natural rightness, but without the definitiveness, less still dogmatism, of notions about 'truth'. There is some affinity of the word with the characteristically Chinese (and equally translation-resistant) concept of *de* (德), often rather stodgily rendered as 'morality' but more properly a somewhat mercurial sense of 'virtue'. Any interpretation of *dao* as a single road is rather undermined by the old form of the character, which shows a man at a crossroads:

When *dao* is used as a verb (the distinction between nouns and verbs is more fluid in Chinese than in English), it has the sense of 'leading' or 'guiding'. But there is nothing constraining or compulsive in that guidance: a person following the *dao* does not need to be kept on course, because that will happen naturally and without effort. The *dao* flows from inner virtue (*de*) like a stream from a spring, and in Confucian tradition a gentleman who has cultivated his *de* does not follow the *dao* from a sense of obligation or duty but simply because that becomes inevitable.

It is because the *dao* is an intentionally and necessarily indistinct concept that water is so apt a metaphor for it. Like water, says Lao Tzu, the *dao* has no fixed shape or form. It flows from an inexhaustible well:

> The *dao* is like a vessel which, though empty,
> May be drawn upon endlessly
> And never needs to be filled.
> So vast and deep
> That it seems to be the very ancestor of all things.

In giving life to everything, the *dao* shares the same nature as water. According to Confucius:

> Water, which extends everywhere and gives everything life without acting, is like virtue. Its stream, which descends downward, twisting and turning but always following the same principle, is like rightness. Its bubbling up, never running dry, is like the *dao*.

As the *Zhuangzi* states, the sage swims as freely in the *dao* as the fish does in water:

> Fish go to one another in water, men go to one another in the *dao* . . .
> Therefore it is said, fish forget about themselves in water; men forget about themselves in the *dao*.

The age of the philosophers

Is it coincidental that the golden ages of Greek and Chinese philosophy occurred both around the fifth to the third century BC? The German philosopher Karl Jaspers argued that the spiritual and philosophical foundations of many of the world's major cultures – not just in Greece and China but also India and the Middle East – were established in the period around 800–200 BC, which he called the Axial Age. Other historians have questioned whether this yoking together of disparate and independent cultures has much meaning. And indeed any significance to the simultaneity of the great philosophers of Greece (Plato, Aristotle) and China (Lao Tzu, Confucius) is rather undermined by the profound differences in what they had to say. The questions that have predominated in Chinese thought are ones that for Western philosophers scarcely seemed like philosophy at all.

For one thing, Chinese philosophy, while at times highly intangible, abstruse and mystical, was never isolated from daily life. One is not surprised to find the influence of Western schemes of thought such as Aristotelianism or Christianity in, say, ethics, art and social relations, but it is rare for those traditions to manifest themselves in the more practical aspects of the quotidian: the Bible doesn't tell you how to build a bridge. The ideas of Confucianism and Daoism, in contrast, were manifested in many diverse spheres of human activity in China. As we shall see, it is possible to speak of 'Confucian engineers' during the Ming and Qing dynasties, and the water metaphors expounded by Mencius to illustrate proper human conduct also furnished guiding principles for hydraulic engineering.

This foundational thinking was constructed on a decidedly unstable political platform. The zenith of Chinese philosophy coincided with social unrest and conflict between states, and the questions that Confucius tried to resolve about personal and public responsibility were lent urgency by the frailty of the status quo and the problems of how to found a robust, lasting state. Conventional histories of China present its Axial era as a steady decline from the eighth century BC – but one shouldn't take that picture too seriously, since there was surely never any 'golden age' from which to fall. Just as the Shang dynasty grew from a loose allegiance of tribes in the Yellow River basin, so too the Zhou dynasty (formally c.1100–222 BC) that succeeded it was no large or stable empire. Yet the Zhou period was undeniably different from the Shang. The latter established a governing hierarchy that went well beyond its tribal origins, and its bronze artefacts are objects of some sophistication and beauty. But the Zhou dynasty was a genuine civilization, with coinage, irrigation, a succession of kings, and most notably, ideas about humanity's place in the cosmos that scholars wrote down for posterity and which shaped all subsequent Chinese thought.

At its peak the 'Western Zhou' kingdom, originally centred in the Wei valley, extended over most of north-central China, including modern Henan, Shanxi and Sha'anxi, and parts of Hebei and Shandong. It deserves to be called an empire, although its unity was lost in the eighth century BC. Earthquakes and eclipses presaged the invasion by a tribe from the north-west who sacked the capital in 771 BC, killed the king and drove the survivors east. The 'Eastern Zhou' sovereign was subsequently little more than a figurehead, lacking any real power over the states that vied for dominance during the next five centuries. From 771 until about 481–475 BC – the so-called Spring and Autumn period* – the kingdom existed as a patchwork of small states which squabbled and fought but never too disastrously. Subsequently these struggles grew more deadly as the states coalesced into just seven major contenders. This so-called Warring States period was a time of legendary conflict between rival kingdoms vying to dominate the country from north of the Yellow River to south of the Yangtze.

---

* The name of this era comes from a text called the *Chunqiu* (*Spring and Autumn Annals*), which describes the history of the state of Lu, in which Confucius is said to have been born, from 722 to 481 BC. Tradition ascribes the work to Confucius himself, although this is surely mistaken. Note that this *Chunqiu* is a distinct, older work than the *Spring and Autumn Annals of Master Lü* mentioned on page 55; the latter's title is an intentional allusion to this classic text.

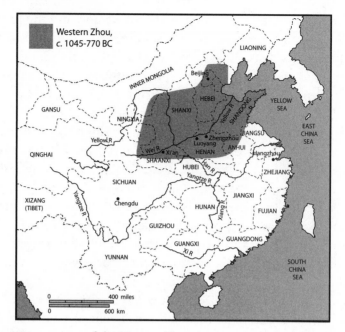

The approximate extent of the Western Zhou empire (superimposed on the modern provinces of China), and opposite, the patchwork of 'Warring States' that followed.

The two most influential schools of philosophical thought in China both arose during the waning of the Spring and Autumn period. While not exactly opposed to one another, Confucianism and Daoism came to be regarded as contrasting philosophies: the first mundane and concerned with how to behave in society, the second mystical and contemplative, considering humankind's situation in the cosmos. Confucianism was public, formal and conformist; Daoism was private, spontaneous and natural. Confucianism established the political and ethical foundations of the state, while Daoism guided the creative, aesthetic and spiritual impulses. This is not altogether a caricature, but nonetheless the two schools were less estranged at their inception than they later became. Indeed, according to Daoist tradition Confucius (551–479 BC) was himself taught by Lao Tzu (Laozi, meaning Old Master). Although modern scholars reject this idea – they are not even sure if Lao Tzu was a real person – the foundational Confucian text, the *Analects*, is composed of rather cryptic short sayings that, like the *Dao de jing* itself, leave considerable latitude for interpretation.

This ambiguity is doubtless in part a reflection of the challenge of translating across time and continents. But it is surely also intentional, for the good Confucian is distinguished by good judgement rather than by adherence to some rigid set of rules. The paradox is that Chinese culture, which has traditionally been regarded (not without justification) as rather tightly confined by principles of etiquette (*li*, 礼), conduct and hierarchy, was built on a philosophical base that is far from prescriptive. Proper conduct was not enforced by some imperious and doctrinaire god, but came from within: from an alignment of the personal with the natural, and from a belief in the capacity for human self-determination.

So while Confucianism, with its notions of civic duty and filial piety, was never far from being a conformist tradition, it did not originally advocate the kind of bureaucratic dogmatism that it later came to represent. Eventually Confucianism became something of a straitjacket for both official and domestic life. In the Neo-Confucianism of the Song period, devised largely by the philosopher Zhu Xi (1130–1200), even heaven has an administrative hierarchy. The supreme deity the Jade Emperor ruled through a graded system of intermediaries, the lowliest of which (the Tudi Gong or Gods of the Earth) were almost like local village magistrates. Within this stringent social framework there was no space for recusancy, and shame fell on anyone who did not respect social convention.

That, however, was later. In its earliest form, Confucianism drew deeply from the well of the *dao*. Although the *Dao de jing* (crudely, *Classic of the Virtues of the Way*), probably written around the sixth century BC, is the best known and arguably the most beautiful of Daoist texts, equally significant in shaping and defining this philosophy is the work called the *Zhuangzi*, attributed to a scholar of that name from the Warring States period in the fourth century BC.

As is characteristic for those classic texts, *Zhuangzi* might or might not have had anything to do with it. (I'll indulge the conceit here.) What little we know of him relies on Sima Qian's account, and it's surely significant that the Han historian chose water metaphors to describe the impact of his writings: they were, he said, like a tidal wave that swamped everything and could not be stemmed, but also so freely flowing that no ruler could encapsulate and harness them. Zhuangzi wasn't exactly a Daoist, however; he was more eclectic in his thinking, at the same time transcendental and possessed of an earthy wit, not to mention a strong Confucian sensibility. He was said to have come from the Meng district of Henan, then within the kingdom of Chu, and that he refused to accept a prestigious state appointment on the grounds that it would stifle his intellectual freedom. King Wei of Chu sent two officials to tell Zhuangzi that he was offered the post of prime minister. They found him fishing in the Pu River, where he responded in this manner:

> I hear that in Chu there is a sacred tortoise which has been dead for three thousand years. His Majesty keeps it wrapped up in a box at the top of the hall in the shrine of his ancestors. Would this tortoise rather be dead, to be honoured as preserved bones? Or would it rather be alive and dragging its tail in the mud?

It would surely prefer to be alive, the officials answered. Whereupon Zhuangzi declared, 'Away with you! I'll drag my tail in the mud.'

Whether or not his love of fishing had anything to do with it, Zhuangzi evidently found water 'good for thinking'. He relates how, when the autumn floods arrived and the Yellow River was swollen to its greatest degree, the Lord of the Yellow River looked across it and felt pride in his ownership of what seemed the most beautiful thing in the world. But then he travelled with the river down to the North Ocean and, confronting the god of this endless expanse of water, recognized his place. And so it is with the *dao*, Zhuangzi tells us: scholars may become puffed up thinking that they know all about it, but before its vastness they are ignorant and insignificant.

His fixation with water prompts Zhuangzi to ask questions about nature that testify to an awakening notion of the hydrological cycle and its meteorological consequences:

> Do the heavens revolve? Does the earth stand still? Do the clouds make rain? Or is it the rain that makes the clouds? What makes it descend so copiously?

Most profoundly, water is what gives humans their life and virtue:

> Man is water, and when the producing elements of male and female unite, liquid flows into forms . . . There is not one of the various things which is not produced through it.

The *Guanzi* elaborates in detail on how water is the source of all vitality, fecundity and organic organization:

> It is accumulated in Heaven and Earth, and stored up in the various things [of the world]. It comes forth in metal and stone, and is concentrated in living creatures. Therefore it is said that water is something spiritual. Being accumulated in plants and trees, their stems gain their orderly progression from it, their flowers obtain their proper number, and their fruits gain their proper measure. The bodies of birds and beasts, through having it, become fat and large; their feathers and hair become luxuriant, and their stripes and markings are made apparent. The reason why creatures can realize their potentialities and grow to the norm is that the inner regulation of their water is in accord.

This idea that water regulates the inner properties of things and creatures leads the author of the *Guanzi* to conclude that the differences in character between the peoples of China's seven major states arise from differences in the qualities of their water sources. In other words, the nature of water defines the morals of citizens not just metaphorically but, as one might say, organically. This makes it all the more vital that a ruler should ensure that the waters of his state are uniform and pure, so that his subjects might live in harmony and virtue:

> It is only he who knows how to rely [on its principles] who can act correctly . . . The sage's transformation of the world arises from solving the problem of water. If water is united, the human heart will be corrected. If water is pure and clean, the heart of the people will readily be unified and desirous of cleanliness. Even when the citizenry's heart is changed, their conduct will not be depraved. So the sage's government does not consist of talking to people and persuading them, family by family. The pivot (or work) is water.

A philosophy of conduct

The worldliness of Confucian philosophy is that of China itself: it shows a commendably practical concern with the business of daily life and the structure of society, rather than with abstract metaphysical questions. This is a philosophy of how to live well, and in his belief in the improvement of society and the value of a sound, open-minded education Confucius seems almost to prefigure the ideals of the European Enlightenment. But the emphasis is on governance and social obligation rather than liberties. Artistic and literary expression in China focused not on deeds of individual heroism like those that were lionized in Greece but on locating humans as social beings within a well-ordered, hierarchical and bureaucratic system that carried specific duties. Individuals were honoured not for martial feats but for their social and moral integrity: they were not mighty warriors but exemplary adminstrators.

We can be reasonably confident that Confucius, in contrast to Lao Tzu, was a real historical personality. The conventional biography says he was born around 551 BC into a humble family in the state of Lu (which is now more or less coincident with Shandong province). He held a minor government position before resigning, disillusioned by the neglectful conduct of the duke. Having established himself as a teacher, he took to wandering throughout China expounding his ideas. The historian John Keay sums up his life with blunt perspicacity: 'Seldom has posterity been so generous; seldom has such a dismal career ultimately been rewarded with such universal esteem.' One might add that seldom has such an influential school of thought sprung from such unpromising soil, for nascent Confucianism showed little prospect of eclipsing its several rivals. That it did so was perhaps due to the ability of its early adherents to monopolize the educational syllabus; there's no better way to capture the orthodoxy.

As I've said, the concept of propriety or etiquette (*li*, 礼) at the core of Confucianism was not at first the rather empty, ritualized formalism that it was apt to become in later times. One might better compare it with Polonius's 'To thine own self be true' in *Hamlet*: the gentleman acts not according to some fixed set of rules but out of genuine, innate sentiment. Personal expression need not be stifled beneath shallow formality, and one's principles were certainly not to be compromised by deference to authority. 'In funerals and cere-monies of mourning', Confucius wrote, 'it is better that the mourners feel true grief, than that they be meticulously correct in every ceremo-nial detail.' It was only among some of his followers that tradition, ceremony and filial duty became obligatory.

At its core, this very pragmatic philosophy was a quest for the *dao*. There was nothing mystical about that: it was simply the path to harmony, contentment and enlightenment, or in other words to a healthy and fulfilling life. There isn't anything fixed or preordained about the *dao*: it depends on circumstances, and consequently one's conduct should do likewise. Water was instructive here in its tendency to find its own level, to achieve poise and repose. 'Perfect balance is found in still waters', Confucius wrote. 'Such waters are an example to us all.'

This was also a political philosophy. State officials, Confucius said, should be selected on merit rather than by birth. He denied that a ruler should be afforded blind devotion and obedience; if he should deviate from the *dao*, then it was incumbent on his administrators and ministers to criticise and challenge him. This principle was often invoked by discontented public servants – albeit at their peril, for emperors found it an easy precept to overlook.

The Confucian king or emperor was thus as much a servant as a ruler of the state: he held his position by virtue rather than by right. This principle was expounded by the scholar Mozi in the fifth to the fourth century BC, who emphasized that the good ruler aligned his will with that of heaven. If he does not, Mozi said, 'Heaven sends down immoderate cold and heat and unseasonable snow, frost, rain and dew . . . Hurricanes and torrential downpours occur repeatedly.' The Confucian disciple Mencius criticized wayward kings harshly; he commended Yao and Shun, who dealt so effectively with the great flood, as historical models for the ideal ruler.

Confucianism owes as much to Mencius (*c*.372–289 BC) as it does to its founder. Like Confucius, Mencius was from the region that is now Shandong. He was born in the minor state of Zou, and became a professional teacher and a government official in the state of Qi. Mencius rooted Confucian thought in an optimistic view of human nature, which he regarded as intrinsically good – an essentially humanistic attitude. In the *Mencius* – compiled by his disciples, and sometimes said to be the real foundation of Confucianism – there is an account of his dispute with a scholar named Gaozi, who claims that 'man's nature is like whirling water':

> If a breach in the pool is made to the east it will flow to the east. If a breach is made to the west it will flow to the west. Man's nature is indifferent to good and evil, just as water is indifferent to east and west.

To this, Mencius responds:

> Water, indeed, is indifferent to the east and west, but is it indifferent to high and low? Man's nature is naturally good just as water naturally flows downward. There is no man without this good nature; neither is there water that does not flow downward. Now you can strike water and cause it to splash upward over your forehead, and by damming and leading it, you can force it uphill. Is this the nature of water? It is the forced circumstances that make it do so. Man can be made to do evil, for his nature can be treated in the same way.

Mencius, says Sarah Allan, can be considered the victor in this debate 'because he had a better understanding of water than Gaozi'. Mencius's remarks also imply, however, that it is unnatural and perhaps even immoral to try to compel water to go against its impulse by 'damming and leading it'.

The First Emperor

The question of human nature was not settled with Mencius's rosy view. For the Confucian writer Xunzi in the third century BC, 'The nature of man is evil; whatever is good in him is the result of acquired training.' This being so, people could not be trusted to think and act for themselves. They needed to be governed by firm authority and to strictly observe principles of proper conduct (Confucian *li*). This *li* is now not something acquired from the *dao* so much as impressed upon one by the prevailing social mores. Xunzi's distorted Confucianism became a justification for dictatorship, and his disciples constructed the philosophy of Legalism that shored up the harsh rule of the Qin dynasty.

At first the brutal military suppression of his opponents by King Zheng of Qin – a kingdom in the Wei valley in what is now Sha'anxi – seemed to promise peace and stability after the turmoil of the Warring States. By defeating the state of Qi in modern Shandong in 221 BC, Zheng eliminated the last of his rivals and became Qin Shi Huangdi ('the First Qin Emperor'), ruler of all China.

Some etymologies trace the name 'China' itself to the Qin (pronounced 'Chin') dynasty, and so you might imagine that it would have a very special status in Chinese history. But the unified state barely outlasted the death of Qin Shi Huangdi himself – four years later it succumbed to a rebellion from which the considerably more durable Han dynasty (206 BC to AD 220) emerged – and it is regarded with little fondness in China today. For the First Emperor, it is said, was a tyrant with 'the voice of a jackal . . . and the heart of a tiger or a wolf', who ruled by brute force. Since much of what we know about the Qin era comes from Sima Qian, who was writing to justify the Han ascendancy, it's possible that the First Emperor gets something of a raw deal. But there is little doubt that he was a despot whose paranoia about the security of his newly forged state drove him to harsh extremes.

All new dynasties were concerned to establish their legitimacy in historical and cosmic terms. As the Eastern Zhou dynasty reached its final stages of dissolution (in truth it had been barely significant for centuries), the philosopher Zou Yan of Qi justified the eclipse of the Zhou rulers in terms of a zodiacal succession of the five elements: earth, fire, water, wood and metal. 'Each of the Five Elements is followed by the one it cannot conquer', he wrote.* Thus the Shang (whose element was metal) had replaced the Xia (wood), and the Zhou (fire) had conquered the Shang. It was a displacement as natural and inevitable as the sequence of animals in the Chinese cycle of years, and presaged by signs in nature: before the rise of the Zhou, 'heaven exhibited fire'. In like fashion, Zou Yan wrote, 'Following Fire there will come Water.' And so it did: water was the emblematic principle of the Qin. Each element had corresponding properties, such as colours: water was black, and that became the hue of Qin Shi Huangdi's livery and banners.

But if water did indeed serve as the 'model' for the state's activities, there was no sign of the softness, reflectiveness and yielding nature celebrated in Daoism and Confucianism. Instead, the Qin Legalists created a philosophical framework for a bellicose, conformist dictatorship in which there was no room for the humanism of Confucius or Lao Tzu. They plundered and rewrote Confucianism and Daoism to support an amoral detachment from the concerns of the people. They celebrated warfare and advocated the slaughter of citizens in conquered lands: 'attack not only their territory but also their people' was the Legalist maxim. The ruler governed not by virtue but by a kind of force called *shi* (势), which connotes both legitimacy or power and the political situation or tendency that the ruler inherits. It is a strict rule of law (*fa*, 法): hence the term Legalism. This philosophy offered a metaphysical justification for the ruthless suppression of critics and opponents. Books that did not suit the Legalist model of statecraft were burnt; many precious documents and works of philosophy and poetry were lost forever. Scholars who failed to conform fared no better. The season of water was winter, a time of darkness and hardship – and the Qin dynasty lived up to that image.

For Qin Shi Huangdi, water, not knowledge, was power; he even renamed the Yellow River 'water of power'. To house his body after

---

* Most of what we know about Zou Yan also comes from Sima Qian. This cosmic cycle and its associated portents became a firm feature of late Han metaphysics, and possibly led to anticipation of the dynasty's downfall.

death he constructed an immense mausoleum near the ancient capital of Xi'an where – according to Sima Qian – his body was placed in a great chamber surrounded by a scaled reconstruction of the entire kingdom: a microcosm that asserted the continuity of the emperor's rule from the afterlife. This landscape was allegedly laced with rivers and lakes that would never dry up because their waters were made of *shui yin*: 'water silver' or mercury, smelted from the cinnabar of Sichuan and Sha'anxi. 'Mercury was used', said Sima Qian, 'to fashion imitations of the hundred rivers, the Yellow River and the Yangtze, and the seas constructed in such a way that they seemed to flow.'*

As well as building his marvellous tomb, the First Emperor compelled his subjects into other colossal feats of engineering. He constructed the Great Wall from existing fragmentary defences on the northern frontier, and had the Lingqu Canal carved out to connect the Yangtze to the Pearl River delta in the south (see page 109). The Qin authorities insisted that individuals must live, work and die for the sake of the state, irrespective of personal interests. They should endure hardship without complaint and be made compliant with threats of punishment. This flatly contradicted the obligations of the virtuous leader according to Confucius, and when a revolt eventually came in 209 BC, Confucians spoke out in favour of the uprising, saying that the emperor had forfeited his throne. The rebel leader, a peasant named Liu Bang, became the first Han Emperor (Gaozu) in 202 BC, whereupon Confucianism was restored as the moral philosophy of the state.

Statecraft and the way of water

'The highest good is like water.' Once you accept this Daoist precept, all else follows: when water is granted moral virtue, then control of water is a source and fount of morality. Water knows its proper course, and in the same way people know instinctively how they should act.

---

* Sima Qian's description was long regarded as hyperbole, but archaeological tests over the past several decades have revealed high concentrations of mercury in the soils piled on top of the unopened tomb in a vast mound. Whether Qin Shi Huangdi really did lie in state amidst mirror-bright representations of his 'waters of power' is something we may never know, for, in contrast to the burial pits containing the 'Terracotta Army' fashioned to protect the emperor's spirit, there are no plans to open up the tomb for examination – and some archaeologists suspect that the chamber has long since collapsed anyway.

The word used for this natural tendency to head towards the rightful place is *gui* (归), which implies a kind of return: a capitulation to what comes naturally. It was used also to denote the way the people were drawn to a just and noble king, and the transition of a woman over to her husband's family after marriage, as well as the flow of a tributary into a river or the river into the sea. In other words, these social norms, so important for stability, were encouraged by making them explicitly a part of the natural order.

That water always finds the lowest level and settles there horizontally was another source of moral instruction. The implication was not, needless to say, that one should lie on a bed in the basement, but that one should find a stable, steady morality. The evenness of standing water was a symbol of justice and an example for how laws should be framed. By seeking the lowest point, water displays its humility. As Lao Tzu put it:

> The reason that the River [Yangtze] and seas rule the hundred valley streams is that they are good at taking a lower position to them . . .
> This is why the sage who wishes to be in authority over the people always humbles himself in his speech.

To be low was to be strong – an idea with implications for gender status. The lower, mightier reaches of a river were regarded as female, and the *Dao de jing* insists that 'the female always overcomes the male by means of stillness'.*

Still water becomes a mirror, and water-filled vessels known as *jian*† were used as ritual mirrors within which the truth could be seen reflected; the dead were sometimes buried with them, so that the reflective surface might provide a gateway between worlds. The mirror-like clarity of still water also supplied the imagery for the state of mind/heart (*xin*, 心) that a sage might attain (in ancient China the heart was considered the thinking organ). As the *Zhuangzi* says:

---

* The *Analects* reassert traditional male supremacy, however, claiming that gentlemen avoid the lower reaches of a river because 'all of the disgusting things in the world are drawn to there'. Bodily metaphors like this abound in Chinese water imagery.
† The character (監) today generally denotes the act of watching over or inspecting. The early forms of *jian* are delightfully graphic:

If water is still, its clarity lights up the hairs of beard and eyebrows, its evenness is plumb with the carpenter's level; the greatest of craftsmen take their standards from it . . . If water, when still, is so clear, then how much more the quintessential spirit. The *xin* of the sage is clear! It is the *jian*-mirror of heaven and earth and the *jing*-mirror [a bronze mirror] of the myriad living things.

Xin also denotes a kind of potentiality: Mencius says that it contains the 'sprouts of virtue', an image that links people with plants in their ability to grow and bloom. The good Confucian ruler channels his virtue (*de*) so that it nourishes the *xin* of his people. And just as the purity of a stream depends on its source, so that of a government depends on its supreme head. (Therein, of course, lies the rub.)

Stillness, softness, placidity, limpid purity: these are also the 'feminine' attributes of water in the West. But it is one of the appealing features of Daoism that the connotations are positive. Precisely because water is adaptable, weak and yielding, seeking the path of least resistance, it is virtually impossible to vanquish: over time, it emerges victorious over the hardest and most refractory substance. By the same token, a leader must also adapt to the vessel – the people – that 'contains' him. This ability to yield, adapt and remain placid is one of the most powerful tenets of Daoism, yet is extremely difficult to achieve. The *Dao de jing* says that:

> There is no one in the world who does not know that the weakness of water can overcome the strong, and its softness, the hard; and yet there are none who can put this into practice.

The passivity demanded of this principle was known as *wu wei*, literally 'doing nothing'. If you found the *dao*, then nothing was precisely what you needed to do: nature would take its natural course, the people would follow a virtuous leader, the world would be harmonious. The wise ruler, said Zhuangzi, does not 'act' but follows a principle of minimal intervention, retreating and letting the people act correctly of their own accord. The Confucian ritualist Xunzi expounded this naturalistic viewpoint, saying that 'if the Way is cultivated without deviation, flood and drought cannot cause a famine'.

With a hands-off emperor, it was all the more imperative that the ministers responsible for daily affairs were of impeccable quality. That was why the meritocratic principle that had elevated Shun under Yao's mythical rule remained so important. During the reign of the Han Emperor Wudi (141–87 BC), the selection of some government officials came to be determined by an examination that tested their knowledge of Confucian classics. This procedure was formalized and made more comprehensive in the Tang dynasty, and it persisted more or less continuously until Qing times in the nineteenth century. In principle these civil service exams were open to anyone; in practice, the cost of education and the luxury of indulging it meant that most officials came from wealthier families. By the year 1000 there were an estimated 130,000 or so government officials, organized along strict hierarchical lines in the kind of complex bureaucracy that China has seemed always to find indispensable. These bureaucrats were well schooled in the Confucian canon: the so-called Five Classics and Four Books, which included the *I Ching, Analects* and *Mencius*. This rather narrow syllabus instructed the officials on how to be a gentleman.

And what exactly did that mean? While in the West nobility was commonly a matter of birthright, the Confucian gentleman (*junzi*, 君子) was not necessarily high-born. He acquired his status from his attitude: he cultivated virtuous, humane behaviour. Anyone could in principle become a gentleman – and while, inevitably, the chance to do so was increased by advantages of birth, nevertheless this egalitarian ideal did occasionally mean that men from humble backgrounds could achieve distinction. (Of course, that says nothing about the dearth of opportunity for women to attain similar rank, which was as lamentable in China as it was in the rest of the world.)

If a gentleman turned to Confucius for a guide to good conduct, Confucius directed him towards water. Why is it, one of his disciples asked him, that a gentleman, on seeing a great river, always gazes at it? Confucius replied by comparing water with the *dao*:

> Where there is a channel to direct it, its noise is like an echoing cry and its fearless advance into a hundred-metre valley, like valour. Used as a level, it is always even, like law. Full, it does not require a ladle, like correctness. Compliant and exploratory, it reaches to the tiniest point, like perceptiveness. That which goes to it and enters into it, is

cleansed and purified, like the transformation of goodness. In twisting around ten thousand times but always going eastward, it is like the will. That is the reason that when a gentleman sees a great river, he will always look upon it.

Perhaps now we can glimpse what Mencius was driving at in his warning, mentioned at the start of the chapter, about the hazards of an exaggerated reputation. Water that collects from rain doesn't get replenished, and so it will eventually dry up. So too if a man's reputation exceeds his natural capabilities and has nothing to constantly feed it, it will eventually become exhausted: he will lack the resources to achieve success. Only if one's actions are fed from the spring of the *dao* will they continue to avail.

There was also a prescription here for the survival of the state. Mencius believed that politics, like nature, observes seasonal recurrences, so that a dynastic change every few centuries was as inevitable as the return of the rainy season each summer. But Mencius wrote for emperors, who had the luxury of considering history on a grand canvas. Lao Tzu's audience, in contrast, during the period when the Eastern Zhou dynasty was disintegrating into chaos, was the prince of a small state struggling simply to survive. The trick, said Lao Tzu, was to imitate water: to yield and weave, to adapt to the circumstances, and to seek the natural course. 'One who is good at overcoming an enemy does not confront him', the sage advised.

Was this, though, sound advice? That became the central dilemma of political life in China for centuries after: should one seize and maintain power by coercion, or relinquish control and follow the *dao*?

4 Channels of Power

How China's Waterways
Shaped its Political Landscape

Inconceivably great are the benefits and the destruction which water can produce.

Sima Qian, second century BC

When the French Jesuit Louis Le Comte arrived in China in 1688 on a mission despatched by Louis XIV, he was deeply impressed by the country's waterways. 'Even if China were not of itself so fruitful a country as I have represented it,' he wrote,

> the canals which are cut through it would alone be sufficient to make it so. But besides their great usefulness in [irrigation] and the way of trade, they add also much beauty to it. They are generally of a clear, deep, and running water, that glides so softly it can scarcely be perceived.

European travellers since Marco Polo often seem more impressed by the waterways of China than by any other feature of the country. 'The Chinese say their country was formerly totally flooded, and that by manual labour they drained the water by cutting it a way through these useful canals', Le Comte continues, somewhat conflating the artificially engineered channels with the work that Yü allegedly conducted on the natural rivers. 'If this is true, I cannot enough admire both the boldness and the industry of their workmen, who have thus made great artificial rivers, and a kind of sea, and as it were created the most fertile plains in the world.'

Today the Grand Canal, the mighty artificial channel reaching from the old Song capital of Hangzhou to Beijing 2,400 kilometres to the north, doesn't always add much beauty to the country. It is a commercial shipping route, like nearly all of China's rivers lined with factories and polluted with toxic heavy metals. But it's not so hard to find glimpses of what Le Comte must have admired. In Suzhou, north of Hangzhou in Jiangsu province, the stone steps leading down to the water from the rear of wooden buildings and the trapezoidal bridges with their half-moon arches look much as they would have done three centuries ago.

As Le Comte discerned, irrigation was one of the key roles of the elaborate network of canals and redirected rivers that was carved into the Chinese landscape. But their most important purpose was as conduits of communication, transport – and power. The elegant junks that plied up and down these waterways carried grain not just to feed the population but to honour the demands of emperors in northern capitals, who decreed that it be shipped from the south in vast quantities for use by the court and armies according to a 'tribute' system allegedly consecrated by Da Yü. The rivers and canals were

A channel off the Grand Canal in Suzhou.

the veins and arteries of the autocratic state, and like the populace they needed to be regimented and contained.

This status was apparent even in ancient times. Advising Duke Huan of Qi about the best location of the state capital, his prime minister Guan Zhong explains (as the fourth-century-BC *Guanzi* attests) that management of water is the key to maintaining social order. There are 'five harmful influences' in nature, he says, including drought and pestilence – but floods are the worst. Uncontrolled water has a symbolic as well as a pragmatic impact: it leads to the breakdown of filial piety and disintegration of social relations.

These are not vague warnings or loose metaphors. The minister explains in proto-scientific detail how rivers get out of control when erosion and deposition at a bend undermine the banks. He goes on to outline the bureaucratic system needed to maintain the waterways, with 'a Water Conservancy Office in each district', staffed with 'men who are experienced in the ways of water'. It is the kind of orderly vision that one might expect to find in baroque-era Holland or Victorian England. The system is quasi-military, and Guan Zhong neglects no details, specifying exactly how many spades, baskets, earth-tampers, planks and carts are required for the job.

A ruler who failed to manage China's waters didn't just risk social decay. He exposed himself to the charge that heaven itself had lost confidence in his capacity to rule. This idea is attributed to the Duke of Zhou, the legendary paragon who helped his brother Wu to found the Zhou dynasty in the eleventh century BC. After his brother's death, the duke acted as regent for his young nephew Cheng, and his impeccable conduct – he resisted the temptation to usurp the throne* – made him the Confucian exemplar of a virtuous leader. Indeed, the writings and thoughts attributed to the duke (many of which are recorded by Sima Qian) are widely regarded as the true origin of Confucianism.

The duke asserted that rulers preside only so long as their good character ensures them a heavenly mandate (*tian ming*). The rulers of the Xia and the Shang (whom, remember, the Zhou displaced) had grown lazy and slothful, and so 'heaven sent down this ruin' on the

---

* By showing a comparable resolve to serve power without trying to seize it, Mao Zedong's faithful right-hand man Zhou Enlai earned the popular nickname the 'Duke of Zhou'. The epithet had a more pejorative tone, however, once the Cultural Revolution discredited anything associated with 'Old China'.

last Shang emperor. Instead, declared the duke, heaven 'chose us [the Zhou] and gave us the decree of Shang, to rule over your many regions'.

There is of course nothing unusual in a leader's claim to have God's approval, and from that time forward rebels and usurpers would routinely justify their uprising by asserting that the ruler had lost the Mandate of Heaven. Conversely, a king endorsed by heaven did not (in theory)need to impose his authority; he would by definition be following the *dao* of the people, and so they would accept him.

Heaven displayed its mandate through natural phenomena. A flood or drought was a sign of divine displeasure. So it was not just the immediate practical consequences of such a calamity that a ruler had to fear, but the conclusion that the people would draw: this must be time for a change of power. In other words, not only might an excess (or dearth) of water trigger social unrest, but tradition more or less *legitimized* such challenges to the prevailing authority. With so little capability to control these natural hazards, a ruler sat on a precarious throne. 'The Mandate of Heaven is not easily [preserved]', the Duke of Zhou noted ruefully. 'Heaven is hard to depend on.'

That's what the regent Wang Mang discovered when he usurped the Han throne. Wang was the nephew of the Grand Empress Dowager Wang Zhengjun, and served as a military commander under the Han Emperor Aidi (7–1 BC). When Aidi died without an heir, the Grand Empress Dowager appointed Wang as regent for the last surviving male descendant of her husband, a nine-year-old boy who became Emperor Ping. During Wang's early regency he cultivated comparisons with the Duke of Zhou himself. But the comparison was misplaced: Wang had designs on power after all. The boy-emperor Ping died five years after – some say, poisoned by Wang. The regent then took it upon himself to pick a successor, a one-year-old boy named Ying – the younger the child, the longer Wang's de facto rule. But in any event, after putting down several rebellions Wang cast aside pretences and declared himself emperor of a new dynasty, the Xin.

This tale sets Wang Mang up as an incorrigible villain, a view encouraged by later Han historians. But Wang made land and tax reforms that testify not only to a capacity for statecraft but also to some genuine good intentions as a ruler. Yet heaven (the Han records claim) turned against him. In AD 11–12 the Yellow River unleashed a terrible flood that shifted its course hundreds of kilometres to the north, creating mass famine and migration that spread social unrest into the valleys of the Huai, Han and Yangtze. Wang's authority never recovered, and by AD 22 a rebel army besieged his court in Chang'an (modern Xi'an). He dreamed up all kinds of crazy technological defences: pontoon bridges supported by swimming horses, human flight, appetite-suppressing pills to counter the lack of food supplies for the army. Nothing worked, and Wang was overthrown the following year. Two years after that, the Han dynasty was restored.

Wang Mang brought the flood calamity on himself, the official annals insist: he was too busy building dykes for protecting the tombs of his ancestors to worry about those protecting the common people. The message was clear: you neglect the rivers at your peril.

Oriental despotism and the hydraulic state

Heaven's mandate, then, depended on controlling the waters. That was achieved – at least, attempted – through the kind of hierarchical, centralized 'mandarin' bureaucracy that is almost stereotypically associated with China. In short, water shaped the political organization of the state. While that idea can be disputed for the loose-knit kingdoms of the Shang and Zhou, since the Qin dynasty it is hard to doubt it.

Was China unique in this regard? The German sociologist Max Weber suggested that rivers were the essential component of political power throughout Asia and the Middle East, including India and Egypt. In these kingdoms, he wrote, 'The water question conditioned the existence of the bureaucracy, the compulsory service of the dependent classes, and the dependence of the subject classes upon the functioning of the bureaucracy of the king.' In this view, it's not enough to say that such a system of centralized rule works best as despotism; rather, despotic rule seems essential and inevitable. That was the argument put forward in the 1950s by the German-American Marxist historian and sinologist Karl Wittfogel. 'Oriental despotism', as he put it, was founded on 'hydraulic civilization'. Of all the challenges created by the natural environment, Wittfogel wrote, 'it was the task imposed by a precarious water situation that stimulated man to develop *hydraulic methods of social control*' (my italics). This idea predates even Weber's speculations. The French philosopher Nicolas Boulanger coined the term 'oriental despotism' in his 1763 book *The Origin and Progress of Despotism*, in which he asserted that the 'revolutions of nature' – by which he meant natural cataclysms such as floods – 'after having destroyed the nations, became subsequently the legislators of renewed society'.

Wittfogel suggested that dominion over the waterways guaranteed control of the means of agrarian production and distribution. In Europe, empires were built by acquiring land; in China that benefitted you little unless you had a means of making it productive. To create irrigation ditches and tend fields and paddies, you needed to mobilize huge numbers of people: it was control of labour that really mattered. Indeed, Karl Marx himself argued that in a civilization as vast, and as lacking in mercantilism, as early China, only the 'interference of a centralizing power of government' could manage so gargantuan a task.

In an age without mechanized resources, hydraulic engineering on the scale needed to keep the mighty rivers of China in check and to exploit their resources demanded the kind of manpower (sometimes female labourers were conscripted too) that could never have been contemplated in Europe, where in the feudal period the idea of 'public works' would have been utterly alien. Mass labour must not only be procured but also coordinated, disciplined and led. The individual matters only insofar as he serves the general 'good' – a servitude that in China was often exercised with total disregard for personal rights, if not in fact for the value of human life itself. The great cathedrals of Christendom were built by workmen who often grumbled about their pay and conditions; but in China, peasant serfs might sometimes be press-ganged into working without any expectation of wages at all, often under conditions of immense danger and hardship. It was easy for the leaders to claim that these efforts were needed for the common good, but there is often little evidence that they had the public interest in mind.

What's more, the mammoth constructions and schemes of a hydraulic state are inclined to demand suppression of any disrupting influence. No institutions or organizations can be allowed to grow strong enough to rival the governmental body politic; no checks to power can be tolerated. When the infrastructure of the state is so dependent on governmental support and control, property rights are weak at best – which meant that a mercantile class like that which eventually came to challenge (and in some cases to replace) the monarchies and aristocracies of Europe could never arise in imperial China. Merchants certainly existed, and they could get rich – but they did so by adapting to a situation in which power was acquired through the state machinery of the civil service. There were, as we shall see, power struggles in the Chinese empires, but they did not have the same anatomy as those that evolved in the West. We can see the legacy of this tradition today, whether in the piratical attitude that, with the state's tacit approval, has long existed to intellectual and material copyright, or in the ruthless expulsion of people from their homes and lands, without right of appeal, in the name of progress.

This isn't to say that the 'hydraulic despotisms' of imperial China exerted total power at all levels of society. Indeed, they had no interest in doing so. As with German fascism or Soviet Stalinism, the leaders recognized that there were diminishing returns to micromanagement: the system gets too expensive and cumbersome. While everything and everyone was ultimately answerable to the heads of state, in practice regional authority was farmed out to local officials, who could get away with a great deal without being held to account. This was a good system from the leaders' point of view. If things went wrong, they could deny responsibility: their task was not to explain failures but to punish them. This abrogation of responsibility meant that corruption could and did flourish. The lessons are still being learnt by the Chinese government today.

Strongly influenced by the deterministic Marxist approach to history (although he became an opponent of Communism after the Second World War), Wittfogel wrote as though his thesis were a mathematical proof, displaying the tendency of Marxist historians to mistake a historical description for an argument that things could not have been otherwise. The Chinese historian Ji Chaoding (Chi Ch'ao-Ting), who studied at Columbia University alongside Wittfogel in the 1930s,* elaborated the hydraulic despotism thesis by arguing that much of Chinese history has revolved around 'key economic areas' of significance

---

* While in the United States, Ji acted as a propagandist and intelligence agent for the Chinese Communists.

both for their agricultural productivity and as communications highways. In both respects, said Ji, the importance of these regions was totally dependent on 'the development of public works for water-control'.

Wittfogel's thesis is generally rejected by historians today, because it fails to capture either the complexity of Chinese water management throughout history or the aims, motives and capabilities of the state. For example, massive state projects were not the only way that water has been successfully managed in China and other regions of Asia. Small-scale irrigation initiatives have also been practised, sometimes effectively, from time immemorial – sometimes with state support, sometimes with regional or local autonomy. Even rather large hydraulic enterprises were occasionally achieved without state support or intervention, such as Qing-era land-reclamation projects in Guangdong and Hunan.

Besides, the image of a monolithic imperial regime ruthlessly imposing its will on an oppressed peasantry, while it might have suited Maoist ideology, is far from the truth. Rulers – even the more despotic of them – wished for a stable and prosperous state, based on an agrarian economy and thriving networks of trade and commerce. They knew that the best way to avoid social unrest and political instability was not through an iron rule of law but through a well-functioning, secure society. They did not in any case have the means to impose their will at every level in every corner of this vast land. Often state intervention in water management represented an attempt to rescue it from corrupt and self-serving local elites and to restore a more equitable arrangement.

So Wittfogel's thesis encourages a simplistic, distorted view of Chinese history, according to which an authoritarian, all-powerful mandarin bureaucracy stymied the ability of the empire to develop a thriving economy and doomed China to stagnation. In this view, merchants had no guarantee of property rights and struggled under the burden of onerous taxation. This picture simply does not tally with the occurrence of several periods of vigorous economic growth, for example during the Song dynasty, the late Ming and the early Qing. It fails to recognize that the state sometimes intervened to encourage industry and growth, not least to stimulate development of more remote areas of the empire.

And yet . . . Wittfogel's thesis has never quite gone away, and this is surely because it acknowledges one indisputable fact about statecraft in China: water is vital to it, and it has always been shaped by patterns of water management, control and access. The image of a hydraulic despotism, says historian Kenneth Pomeranz, 'has long been discredited, but retains some kernels of truth'. Or as political scientist Andrew Mertha puts it, 'the general contours of [Wittfogel's] idea remain powerful and intuitive'.

After all, Wittfogel's association of hydraulic engineering and state-building is enshrined in China's mythology. The German sinologist Hellmut Wilhelm points out that the legend of Yü incorporates the two key features of Chinese agrarian society: regulation of the waterways and the organization of labour that makes this possible. Joseph Needham offered a speculative but intriguing interpretation of a traditional Chinese character for irrigation, *run* (潤). In combining the water radical with the characters for gate (*men*, 門) and king (*wang*, 王), *run* implies water flowing through gates regulated by kings. (Today this character usually means 'to profit', which also tells us something revealing: he who controls the water stands to reap the benefits.) Needham suggested that the failed hydraulic engineers of legend, Gong Gong and Gun, represent rebels who tried to replace feudal kingship with collectivist societies. If that's so, the implication is that without strong leadership a society must fall prey to the kind of chaos that a flood induces. In this sense, water-management issues have long mirrored the challenges and tensions that existed in creating and maintaining a stable Chinese state: the conflict, for example, between the state and the growth of local power centres, such as warlords or wealthy families, that seized control of hydraulic resources.

Underlying Wittfogel's too-tidy thesis, then, is a deeper truth: water management in China has created a *political language*, and it is one that speaks of legitimacy to rule. This language of water continues to inflect Chinese politics today. Critics of the massive engineering projects through which the current single-party state asserts itself must confront the fact that an adherence to these approaches is not simply the result of inertia or hunger for power. In China there remains a deep-rooted and genuine fear that alternative means of governance, whether of water or of society, will bring about social dissolution.

Engineering the environment to serve society is a long-standing tradition in China, and carries a force of moral instruction. Irrigation, for example, was a component of political philosophy; we've seen already how, in Confucian thought, it was used to illustrate norms deemed necessary for the maintenance of social order. Folk tradition provided cautionary tales to reinforce such community-minded behaviour. One of the most familiar of these stories celebrates the power of collective action literally to move mountains. There was, the story goes, once an old man named Yu Gong (loosely 'Mr Simple', also known as 'the Foolish Old Man'), who lived on North Mountain opposite the great peaks Taixing and Wangwu, between the cities of Jizhou and Heyang. The northern flanks of the mountains blocked the road for travellers, who had to make a lengthy detour round them. This vexed Yu Gong, who called his family together and proposed that they clear a way through. The sons and grandsons liked the idea, but his wife objected, saying that the old man had barely enough strength to shift a tiny hillock. In any case, where would they put all the stones and earth? Undeterred, Yu Gong and his sons and grandsons went out with pickaxes and began to hew at the rock and soil. When another neighbour mocked Yu Gong's plan, calling him a fool for spending his last years on this futile scheme, Yu Gong replied that, on the contrary, his success was assured. Certainly, he might die soon – but he has sons, and they have sons, and they too will have sons. His descendants will go on forever, but the mountain will never get any bigger, and so in the end his plan will triumph. To this, the neighbour had no answer.

And truly Yu Gong's case seems unanswerable, for his philosophy has been justified time and again in Chinese history: given enough manpower, anything is possible. But the implication goes deeper. It insists that nature is a malleable thing when confronted by human determination. We can reshape the world to suit ourselves.

Challenging the river gods

'Agriculture', declared the Han Emperor Wudi in 111 BC, 'is the basis of the whole world. Springs and rivers, irrigation ditches and reservoirs make possible the cultivation of the five grains.' But although there are innumerable rivers in the land, the imperial document said, the common people are ignorant about how best to use them. It's therefore the duty of the government to 'cut canals and ditches and build dykes and reservoirs'.

Irrigation, essential for all ancient agrarian civilizations, is particularly complex and onerous in a rice-growing society. Once rice seedlings reach a height of five or six inches, they must be planted in the standing water troughs of paddy fields. The water is drained when the plants are bigger, then brought back when they flower. 'Cultivators of rice build surface tanks and reservoirs to store water, and dykes and sluices to stop its flow', a fourteenth-century document explains. This irrigation apparatus had to be strong enough to withstand seasonal floods and relied solely on gravity-driven flows. It's a wonder that it worked at all.

The irrigated terraced hillsides throughout South East Asia are the work of many generations, and planting out the rice crop – sometimes still done by hand, if the hillslopes hinder mechanized methods – was so labour-intensive that farming families, from children to grandparents, pitched in to help with their neighbours' fields, confident that the favour would be reciprocated. This ancient tradition is thought to have influenced deep-seated patterns of social cooperation and negotiation in rice-growing cultures: it makes the rice-growers of southern China measurably distinct in their psychology from the wheat farmers of the north, more prone to holistic and interdependent patterns of thought.

Irrigation networks in Chinese fields, like this one in Jiangxi province (*above*), are often centuries old. The ditches are used to flood the fields as the rice grows (*right*), and then drain them when they are large enough.

Irrigation was essential for all grain farming, and many of the oldest hydraulic structures were devoted to it. The most ancient public irrigation project now known is the Shi Reservoir in Anhui province, made by the king of Chu in the sixth century BC. In response to the aforementioned decree of the Han Emperor Wudi in 111 BC, the Longgu Canal was constructed to harness the Luo River, a tributary of the Yellow River in what is today Henan. It took ten years to build, but never really worked as hoped. Nonetheless, Sima Qian records the benefits of the Han water projects: 'Wherever the canals passed, the peasants made use of the water. There were tens and hundreds of thousands of ditches – nay, an incalculable number of them – to lead the water from the canals to the fields.'

But this was not the main reason for the canals in the first place, Sima Qian attests. Beyond flood prevention and irrigation, a third motivation for shaping and controlling the waters – and often the one given greatest priority – was their value for transport. It wasn't just grain and other commodities that were borne along the waterways, but armies on journeys of conquest. During the Warring States period, the rival feudal states vied to create and command conduits of power. One of the earliest artificial channels, the Han Gou ('Han-county Conduit') dating from around 486 BC, was made by the kingdom of Wu, which occupied the coastal region of southern China around the Yangtze delta, to create a military navigation route between the Yangtze at present-day Yangzhou and the Huai River. It was extended soon after completion to reach the Yellow River basin. About a century later, engineers of the

Rice paddies in Guangxi province.

Wei kingdom on the middle reaches of the Yellow River built the Hong Gou* to connect the river near the capital of Kaifeng with the Bian and Si rivers, tributaries of the Huai. The canal was initially built primarily for flood control, but in the Han dynasty (when it was called the Bian Canal) it served as a major transport route.

The king of Qin's plans to conquer and unify the warring states included ambitious schemes to engineer the waters. One was the Zheng Guo Canal, near present-day Xi'an in Sha'anxi province, which links the Wei River's northern tributaries the Jing and the Luo to irrigate the plain between them. According to Sima Qian it had a curious history. He says that the engineer after whom the canal is named was sent by the rulers of Han to persuade the Qin ruler to build it. The Han had heard that the Qin were keen on hydraulic engineering, and they figured that such a big project would dissipate their rivals' energies and distract them from thoughts of conquest. Halfway through the construction process, the ruse was uncovered and the Han agent Zheng Guo stood in danger of his life. But he (whether in good faith or to preserve his skin is not clear) argued that, while indeed his original proposal had been a trick, now he could see

---

* It is sometimes suggested that the Hong Gou or 'Wild Goose Canal' dates right back to the sixth century BC, but there is no good evidence for this.

that the canal would confer genuine benefit on the Qin. And it did, creating large tracts of land fit for cultivation and allegedly producing harvests so abundant that the Qin prospered and became mighty enough to overwhelm their rivals.*

The Zheng Guo Canal is one of three projects for which Qin hydraulic engineering is renowned. The second lay far to the south in modern-day Guangxi province: a canal called the Lingqu (often known as the Miracle or Magic Canal), connecting the Xiang River, a tributary of the Yangtze, and the Li River, which eventually reaches the Pearl and thus creates a route all the way to the port of Guangzhou. In this way, exotic produce from the far south of China, such as hardwoods, copper, tin, cinnabar, furs and hides, salt and seafood, could be ferried northwards.

The third of the Qin's great hydraulic schemes enjoys the most enduring fame. Shortly before the unification, a Qin administrator named Li Bing was placed in charge of irrigating the Chengdu Plain in modern-day Sichuan province, which was then the kingdom of Shu. Li Bing tapped the Min River, a tributary (then believed to be the source) of the Yangtze, to turn the plain into a highly productive region.

Li Bing's irrigation scheme was ingenious. To harvest the Min's flow without disrupting it, he split the channel in two. The main channel is shallow and wide, the inner channel deep and swift, so that it can carry most of the flow during the dry season but less when the river is in flood. The inner channel feeds into a network of smaller irrigation ditches controlled by weirs and sluice gates. The division of the flow happens at Guanxian, near the city of Chengdu, where Li Bing constructed an artificial promontory in the middle of the river. Called the Yuzui ('Fish Mouth') in reference to its shape, it was made from bamboo cages filled with rocks and lowered into the river.

The artificial waterways branching from the divided river were also used to drive water wheels for hulling and grinding rice, and to power spinning and weaving machinery. The system was called Dujiangyan, 'All Rivers Weir', a name that now serves for the entire county in this part of Sichuan. To complete it, Li Bing needed to cut away part of the high overlooking cliffs – the base of a peak called Mount Yulei – which he achieved without the aid of blasting (gunpowder was not

---

* All the same, the canal had siltation problems, which required that the channel be recut and new routes be added higher on the Jing during the Han dynasty – a task that left Emperor Wudi with little remaining funding for hydraulic projects.

The Fish Mouth at Dujiangyan in Sichuan, today reinforced with concrete cladding.

yet invented) by heating the rock and then dousing it with water to crack it. Perhaps equally impressive as the feat itself was the fact that Li Bing did not need to conscript forced labour, and there are no records of deaths in the construction, which distinguishes it from several of ancient China's other mega-engineering endeavours. Many of the great hydraulic works from those times needed constant maintenance and repair, and have been either rebuilt or abandoned. But Dujiangyan was designed so well that it is still in use today in close to its original form: it is now a UNESCO World Heritage site, and remains as scenic as it is functional. It can undoubtedly claim to have been of more value to China than the Qin's other legacy, the Great Wall.

The Dujiangyan system was vital to the Qin domination of Shu, and thus to the unification of all China. But the Sichuanese seemed to harbour no grudges, for Li Bing remains one of the few officials from the ferocious Qin era who is still revered. His extraordinary hydraulic scheme made the Chengdu Plain fertile and turned Shu into a flourishing region. Li Bing has joined China's long tradition of 'water heroes', an engineer literally deified, and his taming of the Min has become the stuff of legend. It is said that he overthrew the harsh rule of the river god of Shu, who demanded every year that two virgin brides be sacrificed by drowning in the Min. The families of the victims would be paid a huge sum in compensation, but when Li Bing, as the new prefect of the region, was told that he needed to raise this tribute

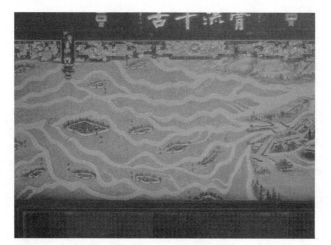

A temple wall at Dujiangyan hosts this detailed illustration of the network of channels that feed off the diverted flow of the Min River.

cash, he insisted that there was no need. For he himself had two daughters who could be the river god's next 'brides'.

When the time came, the two girls were beautifully dressed to prepare them for their awful fate. Li Bing strode up to the throne of the river god on the banks of the Min and poured out the wine as libation. 'Up till now,' he announced, 'I have continued our family line into the ninth generation. Lord of the River, you are a mighty god. Please show your august presence to me, so that I may humbly serve you wine.' But when the god did not appear, Li Bing cried out, 'Lord of the River, you have mocked me, so now I intend to fight you!' He and the god were then transformed into two blue oxen – but Li Bing had taken care to ensure that his officers would recognize which was he, and an archer shot the oxen-god dead. After that, there was no more trouble with the Min.

One theory offers a naturalistic explanation of these events, saying that Li Bing had tumbled a local racket in which unwanted daughters were drowned for financial gain, and called the bluff of the ringleaders by arranging for the theatrical staging of the ox-bout with a deft bit of legerdemain. Whatever the case, such stories show how the prosaic contingencies of hydraulic planning became conflated with China's water mythology. In doing so, legends may confer the legitimacy of tradition on the social and administrative changes involved in managing

Several temples dedicated to Li Bing and his son Er Lang sit on the spectacular cliffs around the Dujiangyan complex.

the waters. Li Bing's son Er Lang, who was also a revered hydraulic engineer, morphed with the Shu irrigation god Guan Kou Er Lang, who is said to have imprisoned a dragon that was flooding the plain. The 'Two Kings' (Er wang) temple to Li Bing and Er Lang still stands at the foot of Mount Yulei at Dujiangyan, where incense sticks release their heady scent in front of statues of the great engineers. On the wall is an exhortation to 'Dig the channel deep, and keep the dykes and spillways low.'

High roads of the Han

The revolt against the harsh rule of the Qin at the end of the third century BC burst forth on many fronts, as former royal families of the old states of Yan, Zhao, Qi, Wu, Chu and Wei rose up against the emperor. Liu Bang, a fugitive who had fled the Qin state in fear of punishment for his failure in an official duty, threw in his lot with the Chu rebels. After the downfall of the Qin, Liu became king of Han in the Sichuan region, and then had to contend with his rival Xiang Yu, ruler of Chu. Liu understood the strategic value of the waterways. If he had not commanded the Wei valley, enabling him to send grain down the river to his armies fighting the Chu forces in Hunan, it isn't clear that he would have emerged the victor. Yet in the end Liu defeated Xiang Yu in 202 BC and became leader of the greater part of a reunified China: the first Han Emperor, Gaozu.

Liu Bang established his new capital city in Chang'an on the Wei River. A grain shortage during the reign of the Han Emperor Wudi made it necessary to transport this commodity up the Yellow River from the fields of the North China Plain to the east. But that was no simple feat. It meant navigating the rocky gorge at Sanmenxia ('Three Gates Gorge'), where the channel is split into three by two islands allegedly created by Da Yü with three great axe strokes, and then towing boats upstream along the Wei at great expense and risk. Much grain and many lives were lost before the minister of agriculture proposed that a canal be dug from the northward bend of the Yellow River, running parallel to the Wei through the Qinling Mountains to the south of the river, to reach Chang'an. This waterway was built in 133 BC with the labour of 20,000–30,000 peasants.

Grain transport to supply armies had begun during the Eastern Zhou era, and both the Qin and Han leaders relied on it. But it was arguably with this creation of a new water route during Wudi's reign that the systematic delivery of 'tribute grain' to the imperial capital began, carried along a water transportation system called the *caoyun* (*cao*, 漕, means 'canal', emphasizing the artificial nature of this network). The grain was delivered to the imperial court to build up a reserve for feeding the officials and the army guarding the northern frontiers, for paying salaries, and for stockpiling against harvest failures. The system remained in place until the Qing dynasty in the nineteenth century.

Grain transport by river junk could be perilous, but carrying it by land, although sometimes done when necessary, was slow, costly and hardly less hair-raising. 'Gallery roads' through river gorges were wooden tracks built onto the sheer sides of the cliff faces: planks resting on stout poles inserted into the rock, generally without even a railing for safety. The grain would be hauled up to the paths by systems of cables and pulleys, then carried by hand along the dizzying narrow walkway. The Chinese people have always seemed extraordinarily sanguine about negotiating these vertiginous routes, although the Tang poet Li Bai made the hazards clear enough:

A walkway along the cliff of Wushan Gorge on the Yangtze. This modern version is blessed with a railing, but traditionally these 'gallery roads' were open on the outside.

> Above, high beacons of rock that turn back the chariot of the sun,
> Below, whirling eddies that meet the clashing torrent and turn it away.

Louis Le Comte gives a visceral account of what the gallery roads felt like to someone less accustomed to them:

> Upon the side of some mountains which are perpendicular and have no shelving they have fixed large beams into them, upon which beams they have made a sort of balcony without rails, which reaches through several mountains in that fashion; those who are not used to these sort of galleries travel over them in a great deal of pain, afraid of some ill accident or other.

Given the dangers of getting grain upriver, it seemed easier in the end to move the capital itself. In the late Han period it was switched east of Chang'an to Luoyang in modern Henan, where it remained during the early Jin (AD 265–311), late Sui (605–14) and much of the Tang (the seventh to the tenth century) dynasties.

A combination of irrigation to boost agriculture and water access for transport enabled new regions to emerge over time as economic centres that might then assert their independence. The irrigation works of the Wei valley in the north made this the power base of the Qin, but the genius of Li Bing created a competing centre in the state of Shu, which remained vigorous throughout the Han era. The later part of that dynasty saw the growth of another key economic region in the lower Yellow River basin towards Shandong. By the third century AD there were thus three powerful territories: Shu in the west, Wei to the north, and Wu in the east-central region. These became three separate kingdoms when the famous Battle of the Red Cliff on the Yangtze (see page 193) sealed the dissolution of the Han dynasty. The Three Kingdoms – all of them in a sense 'water kingdoms' – vied for supremacy for the next six decades.

Taking grain north

From AD 265 until the fifth century, the spell of disunity is tradition-ally patched over by the notion of a fractious and embattled Jin dynasty, although this polity, created by a powerful clan of Wei, could claim to govern a unified China for no more than a decade after it was formed. During this time the area below the Yangtze, called Jiangnan ('South of the River'), was increasingly settled and cultivated, and the south started to become the granary of the north. Water transport was the only practical option for moving grain over such distances. The roads were improved during the Tang and Song periods, but the terrain was difficult: blocked in the south by marshes, lakes and rivers, and obstructed by mountain ranges that forced large detours or made the paths high and hazardous. With river transport, the south's excess of water was turned to advantage: as the proverb had it, 'Go by boat in the south, in the north take a horse.' According to one estimate, in the fifteenth century transporting goods by inland waterway was 30–40% cheaper than taking them by land. Taking them up the coast by sea could be even cheaper in principle, but maritime piracy was rife throughout much of the Chinese medieval period and many vessels were lost to storms.

Joseph Needham suggests that governmental water-management policies in dynastic China prioritized the collection of tribute-grain taxes over the sometimes conflicting needs of flood control and irrigation. Other historians may not agree with that evaluation of the balance, but the importance of water routes is in no doubt. This wasn't just a matter of feeding the population: grain transport was also essential for supplying armies that defended the country's borders against barbarian invasion and suppressed warlords and domestic unrest.

The short-lived Sui dynasty of 581–618 unified both the fragmented nation and (not coincidentally) its north–south grain conduit. The dynasty began like several others: with the opportunistic usurpation of a child emperor. There was nothing at first that seemed likely to distinguish this power grab from the struggles that had kept the country divided for the previous four centuries. Northern and western China were at that time ruled by a dynasty that styled itself the Northern Zhou. When its emperor died in 579, his successor was a libertine with all the trademarks of the degenerate 'last of the line', a man named Yuwen Yun. He too died the following year, and his heir being just seven years old, the empire passed into the regency of Yuwen Yun's father-in-law Yang Jian, the Duke of Sui. By 581 the duke had convincingly marshalled portents showing that heaven's mandate had passed to him. So the young emperor 'abdicated'; he died soon after – from which we might draw our own conclusions – and the duke 'reluctantly' accepted the throne.

The Sui dynasty lasted just thirty-seven years, but during that time it achieved remarkable things. The first emperor was the Duke of Sui, who took the temple name Wen. He and his son Yang (569–618) reorganized the government structure to create the 'Three Departments and Six Ministries', the template for all subsequent imperial administrations. They reformed the coinage and introduced measures to reduce social inequality by improving agricultural production. But they also launched a hydraulic engineering project of such scope and ambition that in the end it destroyed the state.

Emperor Wendi decided once again to move the capital from Chang'an to Luoyang to ensure a reliable connection, via the Yellow River, to the lower Yangtze basin. But for shipping tribute grain, he was not content with the patchy canal network that linked the two great rivers. He ordered the construction of a new system, the 'Great Traffic River' Da Yunhe, known today as the Grand Canal.

Wendi made use of existing canals where possible, including the Hong Gou and Han Gou between the Huai and the Yellow River. He proposed to expand and augment these links to create a continuous passage extending all the way from the coastal city of Hangzhou in modern Zhejiang province, across the Yangtze and the Huai and up to the Yellow River, from where sections would then reach out to both Luoyang and Chang'an. A further branch would extend north from the Yellow River all the way to Beijing.

It was an enterprise of staggering ambition, for which the Sui Emperor mobilized the labour of more than 5 million subjects. There was never a single Grand Canal as such – it was really a system of interconnecting waterways, including rivers and lakes along with old and new sections of canal. South of the Yangtze the Sui engineers built the Jiangnan Canal, running from Hangzhou, via Suzhou, to meet the Yangtze around Jiangdu (today Yangzhou). From here it proceeded north in a section sometimes called the Shanyang Cut,

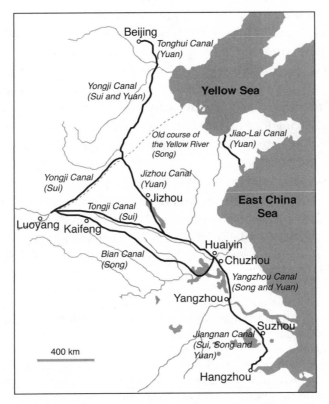

The main water routes that comprise the Grand Canal, indicating the dynasties during which each section was constructed.

which paralleled the old Han Gou to connect to the Huai.* One route north then followed the Bian River, which was extended as the Bian Canal to reach Kaifeng (then called Xunyi) and Luoyang on the Luo River. But in 605 Wendi's son Yang (who had succeeded him the previous year, possibly through patricide) decreed that a new section, the Tongji, be created between Shanyang and the new capital at Luoyang – the precise nature of the route is still disputed, but evidently it connected the Huai and the Yellow rivers. Then a canal called the Yongji progressed north from the Yellow River near Luoyang towards Beijing. This, the longest section at about 1,000 kilometres, also used existing waterways as much as possible, some of which were dredged to improve their navigability. In all, the Sui Grand Canal system covered 2,350 kilometres; it was, Needham writes, 'a main artery to bring tax grain from the economic to the political centre of gravity of the country'.

Yangdi's plans were more or less completed by 609, and the canal was inaugurated in 611 by the progress of an imperial flotilla that is said to have stretched for a hundred kilometres. But the cost was enormous, even before we consider that Yangdi was at the same time building his new capital at Luoyang, carrying out repairs to and extensions of the Great Wall, and paying for a disastrous series of campaigns against the kingdom of Goguryeo in what is today Korea (a conflict that Wendi initiated). To find the funds, he had to impose punishing taxes on the population.

That was not even the worst of it, for the project was accomplished with conscripted labour. According to the Tang-era book *Kai he ji* (*Record of Canal Digging*), to build the Tongji Canal Yangdi conscripted all men – some 3.6 million of them – between the ages of fifteen and fifty, along with a female or older male to make food for every five

---

* The course of this part of the Grand Canal is particularly unclear. There were two old routes from the Yangtze northwards to the Huai, both now referred to as the Han Gou. One, built during the Eastern Han dynasty (AD 25–220), went directly north from Jiangdu to Shanyang, passing through Lake Fanliang. The other, dating right back to the Spring and Autumn period, followed the same southern section but turned north-eastwards from Lake Fanliang to reach Lake Sheyang before heading north-west again to the Huai at Shanyang. Both Wendi and Yangdi built routes from the Yangtze to the Huai that used these old canals, but it isn't clear if they both followed the same path – that is, if Yangdi just rebuilt his father's channels – or if they followed these two distinct routes.

Building canals required huge mobilization of labour from ancient until modern times. The methods did not change so much between the construction of the Grand Canal during the Sui dynasty and this illustration from the mid-nineteenth century of a canal being built at Zhongmou, between Kaifeng and Zhengzhou in Henan.

households, bringing the total to almost 5.5 million workers. That number is surely inflated – the *Kai he ji* is a profoundly unreliable history – but even the more likely estimate of 1 million labourers is vast. The workers were enlisted forcibly (the *Kai he ji* records that anyone caught evading their duty would be decapitated), and they were not well treated; some starved to death. The overseer of the Bian Canal construction was a man named Ma Shumou, to whom posterity gives the attributes of a cartoon villain. Popularly known as Mahu (Ma the Barbarous), he was said to eat each day a steamed two-year-old child, and he became the bogeyman with whom parents threatened their unruly children. His legend, however, testifies to some atrocious memories.

So the Grand Canal accounted for a major part of the financial burden that turned the Sui into a hated dictatorship. Yangdi is traditionally cast as the archetypal decadent 'last ruler', a jealous, stubborn and capricious man who exhausted the patience of his ministers and generals. Rebellions began during yet another of the fruitless assaults on Goguryeo in 613. Li Yuan, the Duke of Tang (a region in modern Shanxi), led an army that overwhelmed the former capital of Chang'an in 617. The rebel duke then declared Yangdi 'retired' and employed the usual tactic of making himself regent over a puppet child-emperor, Yangdi's grandson Yang You. But despite Yangdi's unpopularity, the coup was widely resisted until the emperor himself was killed in 618 during another uprising led by the Sui general Yuwen Huaji. Li Yuan then took the opportunity to make himself emperor: Gaozu, first ruler of the Tang dynasty.

Building the Grand Canal had thus seemed to bankrupt the Sui's heavenly mandate – to the advantage of their successors. As the late Tang writer Pi Rixiu said:

> By dredging the Qi River and Bian River, and by cutting through the Taihang Mountains, [the Sui emperors] had inflicted intolerable sufferings on the people under the Sui. Yet [these projects] have provided endless benefits to the people under the Tang . . . These early accomplishments are monumental. Yet it did not require a single [Tang] labourer to carry a wicker basket, nor a single soldier to chisel through a dangerous place. Is it not true that heaven has greatly benefitted us with the help of the despotic Sui?

It was a common formula: those who came before us were wicked and forfeited the mandate to rule, yet we will gladly receive the fruits of their labours.

In the shadow of the pleasure dome

The foundations of a reliable grain transport system were established during the Tang era (618–907), when typically 130,000 tonnes were moved north every year (the amounts depended, of course, on the harvest). By the Song dynasty (960–1279) this figure had more than doubled, and in the Ming era (1368–1644) it could reach around 450,000 tonnes. Le Comte was amazed by this traffic at the end of the seventeenth century:

> These water-passages, as they call them, are necessary for the transportation of grains and stuffs, which they fetch from the southern

provinces to Peking. There are, if we may give credit to the Chinese, a thousand barques, from eighty to a hundred tons, that make a voyage once a year, all of them freighted for the emperor, without counting those of particular persons, whose number is infinite. When these prodigious fleets set out, one would think that they carry all the tribute of all the kingdoms of the East.

On a good day a riverboat could hope to cover fifteen to twenty-five kilometres. The imperial grain vessels seldom kept the same crew from start to end of a journey – there would be changes along the way, and sometimes the grain might be stored in granaries en route if repairs or bad weather impeded progress. The lives of the river merchants and their captains were a constant procession of misty peaks and water towns, as elegantly conveyed by the Song painter Zhang Zeduan in his famous scroll painting *Along the River During the Qingming Festival*. The Tang poet and administrator Bai Juyi said of a salt-merchant's wife that 'wing and waves are her village, her ship her mansion'.

Le Comte was not the first European to be amazed by the industry that had gone into the Chinese waterways. In the thirteenth century Marco Polo was equally impressed by what Khubilai Khan – the Mongol conqueror of Song China and founder of the Yuan dynasty – had achieved:*

The Great Khan has made very great channels, both broad and deep, from one river to the other and from one lake to the other; and makes the water go through the channels so that they seem a great river; and quite large ships go there with the said grain loaded from this city of Caigu up to the city of Cambaluc [Beijing] in Cathay.

Khubilai Khan was not here just reaping praise that should have gone to his Chinese predecessors. The Yuan rulers, belying the barbaric image of the Mongol nomads, made reforms and developments that

---

* Marco Polo claimed that Khubilai Khan appointed him as a tax inspector at Yangzhou, where the Grand Canal met the Yangtze. Modern scholars still debate how much of his account was invention, and some doubt that he even went to China at all, saying he might just have based his *Travels* on hearsay. The majority view is that the book is authentic in broad outline, though no doubt laced with misunderstandings and exaggerations.

The Song painter Zhang Zeduan's scroll *Along the River During the Qingming Festival* (twelfth century) was much copied; here is a small section of an eighteenth-century version.

transformed China into the worldly, prosperous and sophisticated nation Marco Polo admired – and among them were improvements to the Grand Canal. In Tang times, transport of grain northwards could still be slow and costly: rice harvested in the Yangtze valley might not reach the Yellow River for a year, and significant amounts might be pilfered by local officials along with way. The Tang made changes and repairs to improve the canal, but the Song were less diligent, and some sections suffered from neglect. Khubilai Khan aimed to put that right, and he had the wisdom to employ the capable former Song administrator Wang Jiong, who oversaw some of these renovations.

The Yuan also made a significant extension of the canal's route. It was natural for the Mongol conquerors to set up their capital in the north: Khubilai Khan moved the seat of power from Shangdu in Mongolia (Coleridge's Xanadu) to the Jin dynasty city of Zhongdu (present-day Beijing), which he called Khanbaliq, 'City of the Khan'.* The move meant that the westward leg of the Grand Canal towards Kaifeng and Luoyang – the Bian Canal section connecting to the Yellow River – introduced an unnecessary detour of many hundreds

---

* The Yuan capital is rather encumbered with names. As well as being Marco Polo's Cambaluc, it was also called Dadu, 'Grand Capital'.

of kilometres. It made much more sense to build a short cut extending from the Shanyang Cut north-west across Shandong, to intersect with the Yongji Canal higher up (see the map on page 117). Passing the city of Jizhou, this section is now known as the Jizhou Canal. It was built by a Mongol engineer named Oqruqci and was completed in 1289. As the canal bottom lay about 42 metres *above* the bed of the Yangtze far to the south, keeping it filled with water was a challenge. Here the Yuan engineers were helped by a massive flood of the Yellow River in 1194, which eventually redirected its course so that it flowed in a more south-easterly direction into the Huai basin, raising the water levels there considerably.*

Nonetheless, water supply was always a problem for the Jizhou Canal until it was improved under the Ming in the early fifteenth century. So the water route to Khanbaliq was often inoperable, and the problems were made worse by the unpredictability of the Yellow

---

* This flood contributed to the downfall of the Jurchen or Great Jin dynasty (1115–1234), an incursion of the Manchurian tribe called the Jurchen who seized northern China from the Song in the 1120s. The flood reversed the change of course that had undone Wang Mang's short-lived coup during the Han era (see page 99), and was deemed to imply that the now Sinicized Jurchen invaders had lost heaven's mandate. They yielded to Genghis Khan's army after the Mongol invaders took Kaifeng in 1233.

River itself. The Yuan therefore had to transport grain by sea too, forcing them to collaborate with pirates of doubtful allegiance. When a series of droughts and floods in the 1340s threatened to destabilize their empire, the Yuan embarked on an ambitious scheme to dig a new channel for the Yellow River. But conscription of labour only added to the unrest, and in 1351, the year that the channel was completed, a rebel group called the Red Turbans (see page 141) seized control of the Grand Canal and ultimately overturned the dynasty. Once again China's rulers learnt the hard lesson of attempts to control the waters: you were apt to be damned if you did, and damned if you didn't.

Water towns, canal cities

On either side of the Grand Canal (although surely not for all of its length) Marco Polo describes 'strong and wide embankment roads, upon which the travelling by land also is rendered perfectly convenient'. The roads benefitted the villages and communities that sprang up along the waterway to profit from the passage of the grain fleets.

The Sui Grand Canal began at Hangzhou, then a thriving port and commercial centre on a sea inlet; the harbour silted up during the Ming era. The famous West Lake (Xihu) was originally part of the Qiantang River, but sedimentation of sand spits had turned it into a lagoon by the time of the Qin. In the Tang dynasty it was governed for a time by Bai Juyi (see page 122). He is credited with preserving the West Lake, which had been drying out because of the collapse of a dyke that helped to contain it. In 824 Bai Juyi dredged the lake, raised new dykes, and added a dam to control water flow, so that the lake could provide plentiful irrigation for the surrounding farmland, ensuring Hangzhou's prosperity. The city was already famous for its silk and luxury markets by the time the Song emperors were forced to relocate their capital there in the twelfth century, as the Mongols took over northern China. Hangzhou fell to Khubilai Khan's army in 1276, but the renown, charm and economy of the city was undiminished; when Marco Polo visited towards the end of that century, he conceded that it was 'the most beautiful and magnificent [city] in the world'.

The Sui Emperor Yangdi inspects the Grand Canal at Hangzhou, in an eighteenth-century silk painting.

Downtown Hangzhou today is as generic as any big city in China, but it's not hard, as the sun sets over West Lake, to put yourself back in the days when Bai Juyi would have come to the shores for poetic inspiration. The gathering of elderly men airing their caged birds by the tapering, windowless Baochu Pagoda on a hill overlooking the lake has the feel of a custom centuries old. When, during the Autumn Moon Festival, candles are lit inside the stone pagodas that jut out of the lake to mirror three times the moon's reflection, there is still some magic amidst the couples posing for wedding photos and tourists drinking overpriced Long Jin tea from the surrounding hills.

Few Chinese people will pass up the opportunity to tell the visitor the old maxim that 'Above there is heaven; below there are Suzhou and Hangzhou.' One might suspect that this speaks of a sensibility unfamiliar with Venice or Florence, but nonetheless Suzhou too has

A pavilion on West Lake, Hangzhou.

managed to preserve many of the charms that it possessed when it became a cosmopolitan hub on the Sui Jiangnan Canal. While Hangzhou's lake is regarded (despite its leveed artifice) as a site of natural beauty, Suzhou's attractions are more explicitly human-made – and arguably all the more delightful for that. The exquisitely designed ponds and courtyards of the city's famous gardens, ranging from the intimate Garden of the Master of the Fishing Nets to the sprawling, park-like (and, one can't help thinking, surely misnamed) Humble Administrator's Garden,* are the epitome of Chinese aesthetic refinement. Suzhou, criss-crossed by little bridges over the waterways fed by the Grand Canal, was where officials retired to contemplate the world. For the Italian Jesuit missionary Matteo Ricci in the late sixteenth century, Venice was indeed the obvious point of reference. 'The city is all bridges,' he averred, 'very old but beautifully built.' The city set the standards of artistic style and judgement. According to John Barrow, comptroller to the British envoy Lord Macartney in the late eighteenth century, Suzhou was where one found 'the school

* This administrator, who served the late Han government in the third century AD, was indeed being disingenuous with his humility, saying after his advice had been rebuffed that gardening was all that 'talentless' officials like him were suited for.

The Garden of the Master of the Fishing Nets in Suzhou.

of the greatest artists, the most well-known scholars, the best acrobats', and that 'it rules Chinese taste in matters of fashion and speech'. The Sui Grand Canal made it what it is today, and a fair part of the traffic along the waters carried the silk for which Suzhou became renowned.

But if there is anywhere that you can still hope for a taste of what life on the Grand Canal might have been like before it was plied by oil-drinking barges and crossed by highways on great concrete arches, you will find it in the 'water towns' of Jiangsu and Zhejiang: Xitang, Wuzhen and Tongli, in the region between Shanghai and Suzhou that is so dense with waterways and small lakes that the map looks more like a cross-section through a sponge. True, these places are preserved now for tourists, and in high season you can hardly walk around the network of little canals without being badgered to buy identical merchandise – fake-antique coins, overpriced silk shirts, bamboo toys – but on a quiet day the atmosphere of tranquil contemplation, with not a honking car in sight, gives a sense of why poets and artists sought out the waterside.

Wuzhen water town offers a taste of life on the old canals.

There could be few locations of more strategic importance to China's water transport network than the junction of the Yangtze with the Grand Canal. Two great cities arose here: Zhenjiang on the south bank of the river, and Yangzhou to the north, overseeing a dense network of canals that lead into Gaoyou Lake. The conjoined cities have always been prosperous, the home of merchants, painters, poets and scholars. A group of painters of Yangzhou during the Qing dynasty became known as the Eight Eccentrics because of their individualistic styles and rejection of tradition. The old city is still adorned with canals and bridges.

Entire communities came to depend on the Grand Canal network: they were home to the river workers and offered rest and refreshment for the waterborne traffic. The grain-transport boatmen organized themselves into guilds that, like the Freemasons, took on the aspect of semi-mystical mutual-aid associations. A Buddhist sect called the Luo Zu Jiao spread among the Grand Canal workers around the mid-eighteenth century, and, as such organizations are wont to do, separated into rival networks and factions that sometimes clashed with bloody and even fatal results in the 1820s and 30s.

When the parlous Qing government abandoned its commitment to management of the unruly Yellow River–Huai conservancy in the mid-nineteenth century – the theme of Chapter 6 – the Grand Canal fell into disuse and water transport shifted back to the sea route. This left tens of thousands of boatmen unemployed, along with many more who had supplied services along the canal. These disaffected and disenfranchised men joined rebel groups such as the Red Turbans and the Taipings. Others turned to organized crime, becoming salt smugglers: this lucrative commodity was officially still a government monopoly.

One of the most powerful organizations in the illegal salt trade in Jiangsu during the 1850s and 60s was a secret society with the deceptively benign title of the Friends of the Way of Tranquillity and Purity (Anqing Daoyou), an offshoot of the former boatmen's guilds. In the ensuing decades this sect began to dominate salt trading and other smuggling operations in the ports of the lower Yangtze, fusing with another secret society called the Society of Brothers and Elders (Gelaohui) to create what was in effect a league of bandits. Towards the end of the century it became known as the Green Gang (Qing Bang), and it grew into the major organized-crime syndicate of the Jiangnan/Shanghai region.

Here, then, was where the piratical character of *fin de siècle* Shanghai began: among the disaffected boatmen of the canals. The Green Gang was a decidedly East Asian type of mafia, with strict codes of conduct and honour, quasi-Buddhist initiation ceremonies and supposedly Confucian principles of righteousness, propriety (*li*) and wisdom. Gang members maintained a Robin-Hood-style self-image which harked back to the classic legends of marginalized rebel brotherhoods protecting the poor, as expressed in heroic tales like *Romance of the Three Kingdoms* and *Shui hu zhuan* (*The Water Margin*), set in the marshlands of the Huai basin. And not unlike the case of the American mafia of the 1920s, the Qing government sometimes found it more expedient to form a working relationship with the Green Gang rather than to try to exterminate them. Like the Japanese *yazuka*, meanwhile, they enjoyed a degree of tolerance from the society in which they operated. In the 1920s Chiang Kai-shek would hire Green Gang thugs to disrupt union meetings and strikes.

But they were, of course, gangsters all the same. They ran the notorious brothels, gambling joints and opium dens of Shanghai, they conducted protection rackets, extortion, kidnapping and child trafficking. By the end of the nineteenth century the city was policed mostly by foreign powers, and mainly to protect their own customers and trade interests. The foreigners stayed in their concessions if they knew what was good for them. As for the city proper, 'it is bad form to show any interest in it, and worse to visit it', said the Victorian adventuress Isabella Bird: it identified you as a person of 'odd tendencies'. Outside the colonial district you risked exposure to awful smells, diseases and mistreatment. Yet she did eventually go there, accompanied by an official from the British Consular Service, and recorded that 'I did not take back small-pox.'

When the American journalist Edgar Snow visited Shanghai in the 1930s, he found it 'a fascinating old Sodom and Gomorrah', a 'freak circus' that was 'unbelievably *alive* with all manner of people performing almost every physical and social function in public': perfumed and gowned Chinese ladies, carts of human excrement, peddlers singing out their wares, people playing mah-jong, the scent of opium dens, Englishmen off to cricket or the races, drunken sailors looking for brothels or fights. All of it was there because the Yangtze was there. If Shanghai today takes its status as an economic hub of Asia rather more soberly, still you need not probe too hard beneath the shiny veneer to find surviving pockets of the disreputable and – let's face it – irresistibly alluring old city.

The muddle, squalor, crime and decadence of the Shanghai underworld in the early twentieth century appalled and fascinated Western visitors in equal measure. But for the Chinese governors these conditions reflected the price that was paid when the waterways were neglected: society would decay, social unrest would spread, and perhaps barbarians from outside would seize the advantage. China, it seemed, was always ready to slip back into the chaos of warlord regimes, rebellious cults and squabbling kingdoms – unless the state could impose firm authority on the waters.

5 Voyages of the Eunuch Admiral

How China Explored the World

But before, and before, and ever so long before
Any shape of sailing-craft was known,
The Junk and Dhow had a stern and a bow,
And a mast and a sail of their own – ahoy! alone!
As they crashed across the Oceans on their own!

Rudyard Kipling,
'The Junk and the Dhow'

China, as everyone knows, has been an isolationist nation. While the Europeans went forth on voyages of discovery and conquest, China looked inwards and consolidated its domestic affairs to the point of stagnation.

But all that is quite wrong.

Not only has China for most of its history interacted energetically with its neighbours throughout Asia, and even as far as Africa and Micronesia, but it sometimes did so on a scale that dwarfed the European efforts. When Vasco da Gama rounded the Cape of Good Hope and reached the coast of East Africa in 1498, the indigenous people were unimpressed with his fleet of three galleons. It's not hard to see why: set against the Chinese ships that had sailed to the same coast within living memory, the *São Gabriel* was a pitiful little barque.

The ten-masted Chinese 'treasure ships' (*bao chuan*) of the early Ming dynasty were like floating castles. Some of the largest wooden vessels ever to take to the seas, they weighed 1,500 tonnes to the *São Gabriel*'s 300. And they were heart-stoppingly gorgeous, their broad-beamed hulls as sleek and elegant as a porpoise, carved with dragons' heads painted in bright lacquer and adorned with silk pennants. No one who set eyes on them could have doubted the might of the kingdom that produced them. They could accommodate crews of 1,000 men and their families, who tended vegetable and herb gardens planted on deck.

The treasure ships were aptly named. One scholar wrote of

vessels filled with pearls and precious stones, with eagle-wood and ambergris, with marvellous beasts and birds – unicorns and lions, halcyons and peacocks – with rarities like camphor and gums and essences distilled from roses, together with ornaments such as coral and diverse kinds of gems.

It was the very splendour and scale of Chinese seafaring that makes it all the more perplexing. This was no tentative venture. So what was its purpose? And why did it stop?

A seafaring junk. This vessel is smaller than a treasure ship, but is thought to have been the kind of design on which those mighty ships were based.

The sea in Chinese imagination

It's true that China has never looked upon the sea in quite the way that the Europeans have, and it isn't hard to see why. Europe's coastline is highly fragmented and indented, and in the south it borders an enclosed sea, the Mediterranean, which makes navigation both feasible and desirable: one can reach other lands without too much danger or effort. Even after the invention of the magnetic compass around or before the eleventh century, sailors preferred to navigate within sight of land if at all possible, not just for safety's sake but because one could then be guided by landmarks. In Europe you could get a long way by hugging the coastline; and even if land was not visible, you were never very far from a shore in the Mediterranean. For China, in contrast, the rather featureless east coast pointed out towards the trackless Pacific, alive with dangerous currents, typhoons, and apparently not much else. There were certainly neighbours to the north and south – in contemporary terms, Korea and Japan, Thailand, Vietnam and Malaysia – and China had considerable congress with them, both pacific and military. But to undertake extensive sea voyages required an act of imagination that came much more easily to the Europeans. As the historian John Curtis Perry puts it, 'oceanic experience was not part of the formative rhythms of the many centuries that cut the templates of Chinese civilization'.

China probably also lacked a strong marine tradition simply because its coastline is so much smaller, in proportion to the country's area, than that of Europe. Not many Chinese ever saw the sea. For them, 'water' meant the sweet, nurturing and navigable waters of the great rivers, not the fathomless expanse of the deep. But this isn't to imply that the oceans lacked all interest to the Chinese. A culture so saturated in the imagery of water was not likely to ignore a world entirely given over to that element. The sea was a great and enticing mystery to philosophers, geographers and explorers alike. 'Of all the waters under heaven', says the *Zhuangzi*,

> there is none so great as the ocean. A myriad rivers flow into it without ceasing, yet it is never full, and the Wei-Lü [a kind of vast drain, thought to exist in the Pacific] carries it continually away, yet it is never empty. Springs and autumns cause no change in it; it takes no notice of floods

One of the treasure ships of the Ming fleet, compared with the *Santa Maria* of Christopher Columbus. The scale bar is in feet.

and droughts. Its pre-eminence over the Yangtze and the Yellow River no measures nor numbers can express.

The oceans, like the rivers, were ruled by dragons. There were five of these dragon-kings, one for each of the cardinal points (which in China include the 'centre'). When the dragons are roused, the seas surge. Yet even this mythological menagerie adopts a bureaucratic system of organization, as if the dragons of the rivers and seas were local governors in some great Ministry of Water, complete with coteries of departmental officials. These were, you might say, mythical realms 'with Chinese characteristics'.

Junk art

The Chinese were masterful shipbuilders. The significance and magnitude of the rivers ensured that boats of all sizes were constructed since antiquity. One of the most distinctive features of Chinese vessels was their use of transverse bulkheads: partitions that divided the hulls across their width. In large boats the compartments were generally sealed, isolating the hull in segments and providing a flotation scheme that was resistant to breaches. Europeans did not start using this design until the eighteenth century, and then it was a Chinese import. That

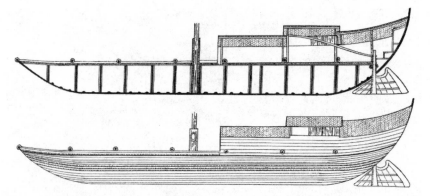

A twenty-eight-metre cargo-carrying river junk from the upper Yangtze, showing the transverse bulkheads of the hull.

was a surprisingly delayed technology transfer, since Marco Polo described the essential principle, and the Venetian explorer Niccolò de' Conti, the first Italian trader on record to have returned from China since Polo, wrote in 1444 that their ships 'are made with chambers after such a sort that if one of them should break, the others may go and finish the voyage'.

Joseph Needham suggests that this system was inspired by the natural compartmentalization of bamboo, the most common construction material in China. Some of the earliest large craft were made from bamboo itself, which can grow up to twenty-five metres high and thirty centimetres across. In the first century AD, bands from Taiwan would cross the South China Sea to raid the coastal villages of mainland China using bamboo rafts that were more or less unsinkable.

'Junks' – the Westernized name for Chinese boats and ships – derives from the character *chuan* (船), in which the first radical (*zhou*, 舟) is still recognizable as a compartmentalized hull. The great sea-ships were called *bo* (舶), while the smallest riverboats, used primarily by fishermen but also for trade transport, were the sampans: relatively flat-bottomed like a punt, and powered by sail or oar. The name comes from *san ban*, meaning three planks, since the hulls really had little more to them than a flat, wide plank on the bottom and two others for the sides.* Somewhat larger river craft of the

---

* The first part of 'sampan' in Mandarin Chinese has now mutated from *san*, three (三), to *shan* (舢), which may serve as an abbreviated term for sampan (舢舨) itself.

Some of the most detailed images of Song-era riverboats appear in Zhang Zeduan's masterpiece *Along the River During the Qingming Festival*.

same design, made from five planks, are called wupans (*wu ban*). One can still find these vessels, almost unchanged in design over centuries, being used by fishermen in China today, winding their way between the iron barges and other industrial ships on the lower reaches of the great rivers.

Chinese riverboats were diverse in form, as the poet Yuan Jue (1266–1327) remarked in evocative images:

> The boats of the Yellow River are like slices of cut melon covered with iron nails for scraping the sandy shallows . . . They come like floating mountains of bundled firewood, scattering before them the boats of Huai and Wu . . . Wu boats are oval-shaped, turtles with tucked-in heads. Families sail them all year round without returning.

Junks were used not only for trade and transport; it seemed that pretty much any affairs conducted on land could be managed on the rivers too. There were junks that housed floating theatres, hotels, restaurants, shops and teahouses, as well as 'singing sampans' from which young women serenaded sailors, and doubtless offered other temptations too.

The second part, meanwhile, means board or plank (*ban,* 板); it is sometimes written, with the boat radical *zhou* attached, as 舨.

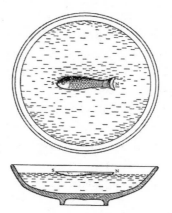

A reconstruction of a floating compass made from a thin leaf of magnetized iron shaped like a fish and floating in a bowl of water, as outlined in the Song manual *Wu jing zong yao* (*Collection of the Most Important Military Techniques,* 1044).

The sails of Chinese boats were usually made not of cloth but of woven bamboo matting, and they hung from the mast stiffened by transverse battens. It is a unique and highly efficient design. One maritime expert in 1906 said of the Chinese junk that

> As an engine for carrying men and his commerce upon the high and stormy seas as well as on vast inland waterways, it is doubtful if any class of vessel is more suited or better adapted to its purpose, and it is certain that for flatness of sail and for handiness, the Chinese rig is unsurpassed.

Chinese boatbuilders were ingenious at modifying the basic designs to suit the purpose. Some river junks were constructed in two articulated halves that could be detached, perhaps as a way of dealing with the unpredictable silting of the waterways: the two halves might separately be able to navigate shallows that the two together could not. Such vessels were also used to devastating effect in aquatic warfare (see page 203).

The earliest sailors of all nations navigated by the stars, along with knowledge of the winds and currents. The use of a compass – probably a magnetized iron needle or sliver floating on water – is attested in China by the late eleventh century, a hundred years before its appearance in the West, and Needham suggests that it might have been used 200 years before that. Chinese nautical map-making was also far in advance of that in the West: what is now known as the Mercator projection was being used in China five centuries before the Flemish cartographer developed this cylindrical format in 1569. There was still no accurate method of determining latitude at that time, but it could be estimated by measuring the altitude of the Pole Star or the Southern Cross constellation above the horizon, which Chinese navigators accomplished by means of a device called a *qianxingban* (a 'star-guide board'). When routes were frequently sailed, crews had the benefit of star maps that would show them the changing configurations of the heavens during the voyage.

Masters of the universe

Extensive sea journeys from China began at least as early as the Qin dynasty. According to Sima Qian, in 219 BC a Daoist priest named Xu Fu was sent by the emperor Qin Shi Huangdi to seek three legendary islands on which magical herbs were said to make the residents immortal. Xu Fu sailed eastwards, returning several years later to say that a dragon or sea mage had promised him the herbs of eternal life if he brought back 'young men of good birth and virgins and workmen of all trades'. You might think this a strange request for a dragon to make; less so, perhaps, for a man wishing to populate and rule his own colony. Be that as it may, Xu Fu left with 3,000 of the said individuals, only to return to request archers to deal with the sharks that encircled the mythical islands. He was granted these too, whereupon he departed once again and never returned. The Qin Emperor was forced to turn instead to his alchemists for an elixir of life, in which they could not oblige their master (they might have actually hastened his demise with their mercury treatments).

Whether or not Sima Qian was just dressing up legend as history, ocean-going ships were certainly being built from at least the second or first century BC, when one is shown on a Han dynasty bowl. The Han sailors traded in South East Asia: Chinese pottery dating from around 45 BC has been found in Sumatra, Java and Borneo, and it has been suggested that the ships might even have reached the Ethiopian coast. The Sui rulers maintained trade and diplomacy with Japan and Korea, and began to exhibit the imperiousness of later Chinese emperors, demanding expressions of fealty and obedience from other nations.

Sea exploration became a rather routine affair during the Tang period, when Chinese ships brought back exotic goods as tribute from all over South East Asia: Indian peacocks, herons, ostriches, zebras, giraffes and black slaves from Africa. Contrary to the common perception that China's coastal economic zones were the invention of Western imperialism, ports thrived in China since at least this time. One of them was Guangzhou, which became every bit as cosmopolitan and polyglot as nineteenth-century Shanghai: Persian and Arab traders mingled with artisans from India and Malay. The Tang emperors Daizong and Dezong in the eighth century commanded ships capable of housing several hundred sailors each. These vessels might be literally home to the crews, who lived, married and died on board, taking residence in dwellings separated by lanes and gardens. The ships could carry a year's supply of grain for their crew; pigs were raised on board. Ships like this crossed vast distances: by Tang times, the Chinese were importing commodities from Africa, including elephant tusks, rhinoceros horns, pearls and incense. In return they sent iron, musk, porcelain, pepper, spices and perhaps silk.

During the Song era, merchants ventured as far as India, and the emperors maintained a redoubtable fleet. The official Lu You saw it on naval exercises on the Yangtze in Hankou:

> There were seven hundred great ships, all of two or three hundred feet in length. They were fitted on top with bulwarks and observation turrets, decked in bright flags and pennants, with their gongs and drums booming as back and forth they breasted the huge waves, as swift as swooping birds. There were tens of thousands of spectators. It was truly one of the great sights of the world.

The scholar Zhou Qufei rhapsodized these seafaring vessels in 1178: 'The ships which sail the southern sea and south of it are like houses. When their sails are spread they are like great clouds in the sky.'

But by the time Zhou was writing, the Song idyll had crumbled. After 1127 the capital Kaifeng and the entire northern part of the kingdom were relinquished to the Jin invaders from the north, and the Song rulers had decamped to the south, setting up the new capital in Hangzhou. Painfully aware of how precarious their state was – they feared attack from Japan and Korea as well as from the Jin – in the 1130s the Southern Song dynasty created what was in effect China's first standing navy. The writer Zhang Yi said in 1131 that now the empire would have to regard the Yangtze and the sea as its Great Wall, and in place of the watchtowers along that barrier of soil and stone these boundaries would need to be guarded by warships. The naval forces grew to 52,000 men over the course of the century; some of the ships built for battling the Jin on the Yangtze were iron-plated and armed with trebuchets and other war machines. A century later the Southern Song navy boasted 600 ships, which gave them control of the East China Sea (see Chapter 7).

But this ocean force could not prevail against the Mongols, who crushed the Southern Song and founded the Yuan dynasty. Accustomed to travel by horse rather than river, the Mongol conquerors nonetheless quickly adapted to the realities of life and war in a river nation, and they paid just as much attention to the navy as the Song had done. When Marco Polo visited Hangzhou under the Yuan at the end of the thirteenth century, he was awed at the marine might on display, while noting that the great ships that 'carry a much greater burden than ours' were once even larger. He claimed that when he left China in 1292 as an imperial envoy, he sailed in a fleet of fourteen immense ships, each bearing 600 men and provisions for two years.

The Yuan ships were made for war and peace. Emboldened by a victorious invasion of Korea, the emperor Khubilai Khan sent envoys to Japan asserting that 'we have become masters of the universe' and demanding that it accept vassal status. When the Japanese military dictatorship (the Shikken) refused the demand, Khubilai Khan launched warships to invade Kyushu in 1274. The failure of that campaign did not deter the Yuan Emperor from subsequently sending a fleet to conquer Java. Again he had no success, but sea power was central to his expansionist ambitions.

Meanwhile, the sea route for transport of tribute grain was revitalized – up the coast from southern China to the northern capital of Khanbaliq. In the early 1280s Khubilai's grain fleet was led by two Chinese pirate chiefs, Zhu Jing and Zhang Xuan, who eventually became admirals in the navy. The early fourteenth century was the heyday of the sea route: by 1329, around 250,000 tonnes were shipped up the coast every year. But there was competition from canal transport, especially after extensive renovation of the Grand Canal. With strong vested interests in the Yuan court, the disputes could be bitter, even deadly. According to one story, the Song minister Wang Jiong, who did much to revive the Grand Canal during the Yuan reign (see page 122), was murdered at sea by supporters of the sea route while on an ambassadorial journey to Japan in 1284. That rivalry was not over by the time the Yuan dynasty gave way to the Ming.

The brilliant dynasty

The man who engineered that takeover seemed at the outset an unlikely candidate for a future emperor, even in a country where dynastic change was often effected by popular revolt. Zhu Yuanzhang came from a very poor peasant family in Anhui, and he lacked any education. He was, moreover, said to be so ugly that he surpassed repulsiveness and merely looked extraordinary. When Zhu was sixteen his family lost their land to a Yangtze flood, and all of them but Zhu and one of his brothers perished in the ensuing plague. He entered a Buddhist monastery where he acquired basic literacy, but after the monastery was destroyed during a local uprising in 1352, Zhu joined a band of rebels and eventually became their commander. They allied themselves with the Red Turbans,* a movement of disaffected agents associated initially with the White Lotus sect popular among workers on the Grand Canal, who opposed the rule of the Mongols.

By 1355 Zhu was the leader of a large army that defeated the Yuan in a river battle at Caishi on the Yangtze, and went on to capture Nanjing the following year: as was so often the case with conflicts in China, military dominance hinged on conquest of the middle Yangtze. Already weakened by infighting, the Yuan then essentially abandoned the south to the Red Turban forces – whereupon the revolutionaries turned on themselves. Zhu commanded one faction, and his main rival was Chen Youliang, whose army occupied the central Yangtze valley. The conflict was settled in a great naval battle (see page 202), and Zhu emerged from it as the prevalent power in southern China.

---

* These rebels didn't really wear red turbans, but rather headbands. The term derives from a poor translation.

In 1367 he declared that the Yuan had lost heaven's mandate because they had 'deserted the norms of conduct', and the following year he announced the beginning of a new dynasty, which he called 'Ming', meaning bright or brilliant. By this time Zhu had defeated the rival warlords in the south, and he marched on the Yuan in the north. There was little opposition: the Yuan Emperor Shundi fled to Inner Mongolia, and the capital Dadu (Beijing) was taken almost without opposition. Zhu renamed it Beiping, 'the North Pacified'.

In later centuries Zhu's victory over the Mongol invaders was cast in nationalist terms as a reassertion of Han Chinese rule. It also marked a shift in the country's centre of gravity, for now the Yangtze, not the Yellow River, was reaffirmed as the locus of economic and political power, and Nanjing replaced Beijing as the capital (although not for long). And yet Zhu, who named himself the Hongwu ('Vastly Martial') Emperor, took over much of the Yuan's administrative framework wholesale, including some of the same Confucian officials (in whose refined presence the untutored emperor felt uncomfortable). He claimed to be re-establishing the institutions of the Tang and Song, but there was in fact a great deal of continuity with the Yuan, who had adopted many Chinese traditions.

Under the Hongwu Emperor, the Ming dynasty had a rather inauspicious start. Insecure and paranoid, he became a tyrant, purging anyone whom he suspected of fomenting disaffection or of mocking his lowly birth and lack of education. There were mass executions almost daily, and his ministers lived in fear of beatings or worse. Officials deemed disrespectful or treacherous wouldn't just be put to death themselves; the same fate would be meted out to their extended families and the junior officials in their command, a sentence that could extend to thousands of poor souls.

Rulers like this do not engender peaceful successions. The Hongwu Emperor died, doubtless to general relief, in 1398, and the throne passed to his grandson Zhu Yunwen, whose father had died six years earlier. Zhu Yunwen pointedly adopted the name Jianwen, meaning 'Establishing Civility': a quality in short supply during his grandfather's reign. But the succession was contested by his uncle Zhu Di, the fourth son of Zhu Yuanzhang, who had been made Prince of Yan (roughly modern Hebei province) in northern China. So began a civil war known as the Jingnan Rebellion.

It lasted for three years, but Zhu Di eventually triumphed when in the summer of 1402 he seized Nanjing. He proclaimed himself the third Ming Emperor, Yongle: the name means 'Perpetual Happiness', but at first there did not seem to be much prospect of that. Scholars and officials in Nanjing who denied Zhu Di legitimacy suffered the fate his father had introduced, being slaughtered along with their extended families: grandparents, parents, uncles and aunts, siblings, children and grandchildren. When one scholar was executed, so were all his unfortunate former students. And Zhu Di reintroduced the barbaric 'death by a thousand cuts', whereby victims were killed by being gradually sliced up with razor-sharp knives.

But having established his power and authority, the Yongle Emperor placed his father's ad hoc dynasty on solid foundations and began to transform it into one of the most glorious and productive in Chinese history. His vision extended to the world beyond the country's borders. No sooner had he taken the throne than he began to build a great fleet, mostly in the shipyards of Longjiang near Nanjing. Between 1404 and 1407, more than 1,600 vessels were assembled.

The Ming voyages transported not only sailors and ambassadors but a microcosm of Chinese society. There were court officials, all of them eunuchs, who supervised the military forces. Protocol was enforced by officials from the Ministry of Rites, and there were physicians and herbalists, translators, astrologers and geomancers, craftsmen, family members, and seamen and soldiers made up largely of criminals. They took with them silk, cotton, ironware, salt, hemp, tea, wine, oil, candles, porcelain: all the riches of China, beyond compare in all the lands around.

Such an extraordinary enterprise needed an extraordinary leader. It had one, and like the first Ming Emperor he was an unlikely candidate.

Adventures in Asia

To consolidate his empire, the Hongwu Emperor had to overcome Mongol Yuan resistance in the remote territories to the north and south-west. Yunnan province remained under the rule of a Mongol prince until the Ming conquered it in 1381–2. On that campaign they took many prisoners, including a young boy named Ma He. It seems probable that he was not of Han Chinese stock but was a Muslim from one of Yunnan's many ethnic minorities. Ma was selected to become a court eunuch, and so he was castrated – the procedure was crude, savage and extremely dangerous, generally involving removal of all the genitals – and sent to join the imperial household in Beijing. There he was renamed Zheng He, and he became a close companion of the young prince Zhu Di, later Prince of Yan, who helped to defend the northern frontier of the Ming empire. In these northern campaigns Zheng He proved himself as a resourceful military commander. He became known by the nickname San Bao, 'Three Treasures', a reference to the three jewels or treasures in which Buddhists find refuge and guidance: the Buddha, the teaching or *dharma*, and the community or *sangha*. Despite his ancestry, Zheng He seems to have been a devoted Buddhist all his life.

He served as a general during Zhu Di's struggle with his nephew the Jianwen Emperor, and was so capable that, after claiming the throne, the Yongle Emperor selected him as the admiral of his new fleet, despite the fact that Zheng He lacked any previous naval experience. He doesn't fit the usual image of a court eunuch: Zheng He was said to be seven feet tall, with a girth to match, and his voice was 'as loud as a huge bell'. He outlived the emperor and commanded all of the great Ming sea voyages.

'Bearing vast amounts of gold and other treasures', write the authors of the *Li dai tong jian ji lan* (*Essentials of the Comprehensive Mirror of History*) in 1767,

> and with a force of more than 37,000 soldiers under their command, [the Ming] built great ships, sixty-two in number, and set sail from Liujiagang in the prefecture of Suzhou, whence they proceeded . . . on voyages throughout the western seas. Here they made known the proclamations of the Son of Heaven, and spread abroad the knowledge of his majesty and virtue.

At its peak, the fleet contained 3,800 ships, including around 400 warships and 250–300 treasure ships, and carried a host of around 27,000 men, mostly soldiers. Each of these great *bao chuan* was accompanied by a little flotilla of support vessels, including supply ships, combat ships and patrol boats. This vast armada is comparable in size to the entire British, French and Spanish fleets at the Battle of Trafalgar. In its prime, Needham says, 'the Ming navy probably outclassed that of any other Asian nation at any time in history, and would have been more than a match of that of any contemporary European State or even a combination of them.'

It's not clear exactly how big the grandest of these naval vessels were, but the dimensions given in old texts, once dismissed as exaggeration, are now taken seriously – in which case the treasure ships were probably the largest wooden seagoing craft ever made. Some estimates give the dimensions as around 135 metres long and 55 metres wide – which is much larger than the European ships of the day, and fits with the scale implied by a rudder post of a Ming treasure ship unearthed in 1962 from the former Longjiang shipyards. The cost of building and maintaining them was enormous. The historian Edward Dreyer estimates that the Yongle Emperor must have spent around 5–15 million piculs (one picul is equivalent to 107 litres of grain) on constructing the treasure ships, which compares with a likely annual grain tax receipt (accounting for 90% of all state revenue) of 30 million piculs.

The first naval expedition, with perhaps just sixty-two or so treasure ships, set out in 1405. It left the Chinese port of Liujiagang (today Liuhe) in the Yangtze estuary, stopped in Changle on the coast of Fujian province, then headed to Champa (modern Vietnam), up the straits of Malacca to northern Sumatra, and across the Indian Ocean to Ceylon and Calicut in India (modern-day Kozhikode in Kerala). In Calicut, Zheng He traded for spices such as cardamom and cinnamon, then sailed home.

The Yongle Emperor was evidently pleased with the tributes he received, for no sooner did his fleet return in 1407 than it was sent out again. Seven times in all Zheng He set forth, and until 1422 he was at sea for much of the time. On the return trip of the first voyage he fought and defeated the fearsome pirate chief Chen Zuyi, a native of Guangdong, off the coast of Sumatra, capturing Chen and bringing him back to Nanjing to be executed. Some countries extended a warm reception and offered rich tributes and commodities. In Siam, Zheng He traded in hardwoods, aloes, incense, ivory and precious feathers; from other nations he brought back exotic medicinal materials such as sulphur, ground rhinoceros horn and deer antler, frankincense, myrrh and camphor. In Calicut the admiral heard the story of a prophet called Mouxie – a transliteration of Moses – who, it was said, had established a religious cult to the west.

Elsewhere the locals proved hostile. During the third voyage (1409–11) Zheng He fought and defeated the troublesome Vira Alakesvara of the powerful Alagakkonara trading family, who had set himself up as the ruler of Ceylon. Alakesvara, who seemed to have been committing acts of piracy that made the sea route to India unsafe, refused to pay tribute to the Ming Emperor and rejected the stone tablet that the emperor sent bearing inscriptions honouring the Buddha. Instead he attacked Zheng He's fleet with a large army. After a bloody struggle in which the Ming troops engaged with the Ceylonese on land, Zheng He triumphed and brought the pretender back to Nanjing in chains. For some unknown reason he was spared execution and released, but the old Sinhalese rulers were reinstated in Ceylon.

Zheng He had to engage in military action again on the fourth voyage of 1412–14, when a rebel in Sumatra (Semudera) named Sekander, enraged that he was not afforded gifts that the Ming fleet conferred on the prince of that region, launched an ill-considered assault on the Chinese. Again, Zheng He's forces captured the enemy leader alive. On that voyage the Ming fleet sailed as far as the Persian sultanate of Hormuz, where the Arabian Sea meets the Persian Gulf. The fifth voyage in 1417–19 took Zheng He to Mogadishu and Brava (Barawa) on the east coast of Africa, where he was given camels and ostriches.

On fair days the fleet might cover sixty sea miles; when the storms raged, the sailors would call out to the celestial consort Tianfei to beg for her protection. 'We have traversed more than one hundred thousand *li** of immense water spaces', declares a stele (stone monument) at Changle dated to 1432 and allegedly bearing Zheng He's own words, composed to mark his final journey:

> [We] have beheld in the ocean huge waves like mountains rising sky-high. We have set eyes on barbarian regions far away hidden in a blue transparency of light vapours, while our sails, loftily unfurled like

---

* The 'one hundred thousand' will not have been meant too literally – any rounded decentile large number tended to stand in for 'a great many'.

clouds, day and night continued their course with starry speed, breasting the savage waves as if we were treading a public thoroughfare.

The *Li dai tong jian ji lan* attests that Zheng He's exploits 'were such as no eunuch before him, from the days of old, had equalled'. It's hard to argue with that.

The attitude of the Chinese envoys in South East Asia was one of lordly patronage and entitlement. They recognized foreign sovereignty – indeed, the Ming Emperor seemed to feel it was his right to confer it, as he did for the kingdom of Malacca, founded on the coast of Malaysia by a Sumatran rebel named Parameswara. The emperor dispatched Zheng He with an official seal declaring that Malacca was a legitimate nation, in the belief that this would bring stability to the region. But at the same time the Chinese expected their hosts to declare fealty to the emperor as a kind of de facto governor of all of Asia. When they refused, as Alakesvara of Ceylon did, the response was harsh.

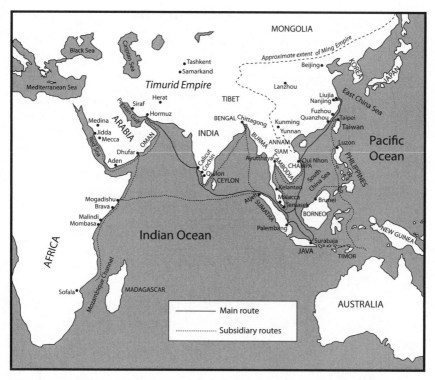

The routes of Zheng He's voyages.

The sight of the Ming Emperor's astonishing power and wealth represented in the elaborately carved and painted treasure ships helped to bring a stream of ambassadors to the imperial court to pay tribute. They came from Bengal, Calicut, Cochin (in modern Kerala) and Java, even from Japan, where the shogun Yoshimitsu was favourably disposed to Chinese culture. Parameswara of Malacca is said to have given the emperor a pair of eyeglasses, perhaps from Venice – which for once was something one really couldn't find in China. The imperial menagerie was filled with strange animals, including lions, leopards, Arabic horses and a Kenyan giraffe, which the Chinese took to be the mythical creature called a *qilin* – an expedient interpretation, given that the appearance of a *qilin* was said to augur a period of prosperity and peace. In the capital, visiting ambassadors would be treated to an impressive display, with banquets, archery contests and horse races.

How far did the Ming go?

It is hardly surprising that the scale of the Ming expeditions has attracted far-fetched claims about their extent. The British historian and sinologist Charles Fitzgerald argued that they reached Australia, in part thanks to the discovery of an ancient Chinese Daoist statue buried among the roots of a banyan tree in Port Darwin. The evidence for the claim is tentative at best, but Needham pronounced it 'not implausible'.

The idea that they went to the Americas is far less so. This notion is best known through the book *1421: The Year China Discovered the World* (2002) by the British submarine officer Gavin Menzies, in which he asserted that in that year Zheng He's fleets rounded the Cape of Good Hope, passed over the South Atlantic, and visited South America before returning to China across the Pacific. Menzies based his claim on genetic studies, alleged archaeological evidence (including the wrecks of Chinese ships off the Bimini Islands of the Bahamas) and indications on Chinese maps. The hypothesis is almost universally dismissed by historians, one international group of experts calling the book 'a work of sheer fiction presented as revisionist history'. None of this deterred Menzies from claiming six years later that China 'sailed to Italy and ignited the Renaissance', a story equally lacking (other historians say) any credible evidence to support it.

Menzies' revision of American history was nothing new. As early as the eighteenth century the French sinologist (such scholars existed even then) Joseph de Guignes claimed that the Chinese had sailed to Alaska and California in AD 458. Such fantasies found adherents throughout the nineteenth century, and even Joseph Needham was prepared to state that 'a mountain of evidence is accumulating that between the 7th century [BC] and the 16th century [AD] . . . occasional visits of Asian people [to the Americas] took place, bringing with them a multitude of culture traits, art motifs and material objects.' There are, for example, suggestions that Mayan art of Central America in the fifth century AD shows Buddhist influences. These ideas seemed all the more likely after Thor Heyerdahl sailed on the *Kon-Tiki* raft from South America to Polynesia in 1947, proving that the voyage was possible on a simple vessel. Might the first Qin Emperor's envoy Xu Fu have made his way across the Pacific to America, Needham wonders? Or might he, more prosaically, have just set up home in Japan? The historical record stays silent, as it so often does.

Discovery or empire?

The voyages of Zheng He were unlike anything the world had seen before, and were unmatched for centuries after. But as to what their purpose was, opinion remains divided. Unlike the Europeans, in creating this astonishing navy the Chinese do not seem to have been motivated by thirst for foreign conquest or by missionary zeal, and even trade did not have the same imperative. The patrons of the Portuguese and Spanish explorers anticipated recouping their expenditure from trade revenues; not so the Yongle Emperor.

With this in mind, Needham presented a benign vision of Chinese maritime exploration. The great sea voyages from the Tang to the Ming, he said, were prompted primarily by curiosity about the wider world: a wish to explore for its own sake. They constituted 'an urbane but systematic tour of inspection of the known world'. Of course, trade did occur, and it flattered the emperors to be given gifts and homage from distant rulers – but the adventures of Zheng He were untainted by cultural or political imperialism. 'In Arabia', Needham writes, 'they conversed in the tongue of the Prophet and recalled the mosques of Yunnan, in India they presented offerings to Hindu temples, and venerated the traces of the Buddha in Ceylon.' The explorations were 'calm and pacific, unencumbered by a heritage of enmities; generous (up to a point), menacing no man's livelihood; tolerant, if more than a shade patronizing; in panoply of arms, yet conquering no colonies and setting up no strongholds.'

This has become the official line in modern China, where Zheng He's journeys are celebrated as amiable and inquisitive, in contrast to the rapacious exploitation practised by the Europeans. The voyages have been promoted by the state as 'excellent materials for conducting patriotic education for the Chinese nation'. Interest in Zheng He, which languished for hundreds of years, was revitalized in the early twentieth century as a source of national pride following humiliation at the hands of the Europeans and Japanese. Those defeats – in the Opium Wars and the First Sino-Japanese War of 1894–5 – were attributed in considerable measure to the weakness of China's sea forces, and so it seemed opportune to hark back to a moment in history when China had dominated the seas. That narrative was, however, adjusted to suit a Western template: these were voyages of discovery to rival those of Columbus, Magellan and Vasco da Gama. That's how Zheng He was presented in an article of 1905 by the scholar and modernizer Liang Qichao, titled 'Zheng He: A Great Navigator of Our Homeland'.

Isn't this the perfect historical precedent for China's present-day policy of peaceful expansion? In a speech at Harvard University in 1997, President Jiang Zemin praised Zheng He as the model for the international dissemination of Chinese culture, and much the same view was articulated on the 600th anniversary of the first Ming voyage in 2005. Andrew Erickson of the US Navy's China Maritime Studies Institute suggests that 'Zheng He's legacy now serves as a sounding board for Chinese maritime ideology and conceptions of maritime moral exceptionalism.'

But were the Ming voyages really so congenial? It now seems more likely that they should be regarded as a form of 'power projection': a means for China to extend its influence throughout the world without military conquest. The early Ming emperors were in fact rather aggressively expansionist, as evident from the way the Hongwu Emperor forced his way into Yunnan – a separate kingdom called Nanzhao or Dali until the Yuan conquered it – in 1381–2, deposing the Mongol prince who still ruled it and settling it with Chinese families. In 1406, on the pretext of restoring the deposed former ruler of northern Vietnam (the kingdom called Dai Viet, or Annam in China), the Yongle Emperor sent a military force into the country to fight the usurper Ho Quy Ly. Ho was captured in 1407 and brought as a prisoner to the court in Nanjing, and the Ming Emperor annexed the territory, calling it Jiaozhi province. It remained under Chinese rule until a revolt in 1427.

On the whole, colonization of countries accessible only by sea was regarded as too impractical. But the places visited by Zheng He were expected to show a submissive attitude towards the emperor. That's hinted in the stele inscription at Changle, which announces that the admiral was commanded 'to go to the [foreigners'] countries and confer presents on them so as to transform them by displaying our power while treating distant peoples with kindness'. The Qing-era work *Li dai tong jian ji lan* certainly makes it sound as though, for all that Zheng He 'bestowed gifts upon the kings and rulers', he expected in return to receive an unconditional expression of allegiance to the emperor. 'Those who refused submission they overawed by the show of armed might', the document says. 'Every country became obedient to the imperial commands' – through coercion if necessary. A show of might could develop into invasion and conquest, as Vira Alakesvara of Ceylon discovered to his cost. There was little a foreign ruler could do to resist the imperial fleet, which carried more soldiers than the entire male population of most of the ports at which it called. These were 'military missions with strategic aims', says the historian Geoff Wade. Edward Dreyer, who re-examined the Zheng He journeys in 2007, points out that if the voyages rarely led to actual combat, that was because the Ming armada 'was frightening enough that it seldom needed to fight'. While the *Ming shi*, a history of the Ming compiled by Qing historians in 1739, is hardly an unbiased

assessment of what the Qing dynasty's predecessors achieved, it anticipated these modern reassessments in saying that the fleet 'went in
succession to the various foreign countries, proclaiming the edicts of
the Son of Heaven and giving gifts to their rulers and chieftains . . .
Those who did not submit were pacified by force.' Dreyer argues
that the Ming voyages are thus best seen as 'power projection . . . to
force the states of South East Asia and the Indian Ocean to acknowledge the power and majesty of Ming China and its emperor'. Wade
adds that they were 'intended to create legitimacy for the usurping
emperor, display the might of the Ming, bring known polities to
demonstrated submission to the Ming and thereby achieve a *pax Ming*
throughout the known world and collect treasures for the court'. The
Yongle Emperor might have anticipated Lord Palmerston by four
centuries with a form of gunboat diplomacy that was aimed not so
much at colonization as such but at dominating trade routes while
also securing the loyalty and tribute of foreign nations. As we saw
earlier, China was conducting sea trade at least since the Song era,
yet Indians and Arabs still controlled most mercantile activity in Asia
until the fourteenth century. The early Ming emperors made trade
an official component of state expansionism: not only did the Hongwu
Emperor ban foreign traders from setting up residence in Chinese
ports, but Chinese trade itself became a state monopoly, forbidden
to private merchants.

Was this, then, really a kind of colonialism? The meaning of the
word is contended, and certainly Ming China was not pursuing the same
sort of imperialism as European nations; the fealty demanded by the
Son of Heaven was more a matter of cosmic ritual rather than mundane
governance. But if the Ming emperors had persisted in extending their
reach throughout Asia and East Africa, a confrontation with Western
colonialism would surely have arrived sooner or later. That gives Zheng
He's legacy contemporary resonance. While one can't draw too close
a parallel between Ming power projection and the 'soft power' of
contemporary China exercised through foreign trade and investment,
President Xi Jinping's talk of a 'Twenty-first century maritime Silk
Road' expresses a keen awareness of the historical precedents.

A quite different 'explanation' for the Ming voyages was offered
by a contemporaneous biographer of Zheng He, who claimed
that Zhu Di remained concerned that his nephew Zhu Yunwen,

the Jianwen Emperor, had not after all (as is almost certainly the case) perished in the siege of Nanjing in which he was overthrown, but had escaped. The Yongle Emperor was allegedly tormented by these thoughts. When an official had had the temerity to raise the subject, seeming to question the emperor's right to rule, his tongue was cut out. But with the blood flowing from his disfigured mouth, the official wrote on the palace floor in Nanjing: 'Where is King Cheng?' (this being another name of the Jianwen Emperor). These characters could not be washed away and even glowed in the dark – and that was why the Yongle Emperor felt compelled to relocate the capital to Beijing (then called Shuntian). In this view, Zheng He was sent out into the world to search for the deposed relative. It's a good story – but is best regarded as a fable symbolizing the emperor's genuine concern to establish the legitimacy of his rule.

The real reason why the Yongle Emperor sought to move the capital back north was so as to be able to oversee the military defence of the northern frontier against Mongol invaders. He began this process almost as soon as he came to power in 1403, constructing the lavish Forbidden City and relocating 10,000 households – mostly peasants from Shanxi province – to keep him company: an early intimation of the Chinese government's assumption that lives can be shifted around like the pieces on a *xiangqi* board. Immense quantities of tribute grain needed to be shipped to the northern capital – the quantities doubled compared to the Hongwu era (when the grain was used mostly by the army in Liaoning), reaching well over a million tonnes a year. But the ocean route was too prone to shipwreck and piracy to be entrusted with such an important task, and so Yongle ordered that the neglected Grand Canal be restored. The waterway was fit again for navigation around 1411, and sea transport of grain was abolished soon after. The new capital was not formally inaugurated until 1420.

Yet the early days of the northern court did not augur well. First, hearing of a plot to assassinate him, in 1421 the emperor executed hundreds of eunuchs and concubines suspected of being complicit. That spring he fell badly from a horse given to him in tribute. Then the Forbidden City suffered bad fire damage after being struck by lightning. Things never improved much after that, and the Yongle Emperor ended his days as suspicious and unpredictably vicious as his father. Zheng He was away on a mission to Palembang in Sumatra when the emperor died in mid-1424. When the admiral returned to his homeland, he found that things had changed.

End of an era

The Ming voyages resembled many other immense Chinese engineering projects in so far as they pushed the prevailing technologies (and state finances) to their limits – and because, once officially sanctioned, they acquired their own momentum. But just as a single decree could launch them, so too they could be terminated at a stroke.

The Confucian officials in the imperial court had long insisted that these sea voyages were a pointless drain on finances and resources that would be better directed towards domestic water conservancy and agricultural projects. There was nothing, they said, that China could usefully learn from the barbarian nations, and in any event it was not seemly to revel in luxury goods like those that the treasure ships brought back. These officials were also attempting to wrest power from the eunuchs, who supported the missions. When Yongle died in 1424, his son and successor, Zhu Gaozhi, the Hongxi Emperor, sided with this anti-maritime party. In September he commanded that all voyages of the treasure ships were to be stopped and the fleet was to be recalled to Nanjing. Zheng He was removed from his post, although he was placed in control of the army.

The Hongxi Emperor's decision wasn't exactly a surprise. Already by the time of the sixth voyage in 1421–2, his father was forced to recognize that the cost was excessive, especially while he was at the same moment conducting military campaigns in Vietnam (Champa) and Mongolia and building his new capital. That voyage was a last gasp (or almost), after which the emperor agreed to suspend the sea missions. His son merely made this ban permanent.

The seafarers gained temporary respite when Hongxi died a year later, since his son and heir, Zhu Zhanji, called the Xuande Emperor, took their side. With China's prestige apparently on the wane and foreign countries less inclined to send tribute, in 1430 the new emperor issued a decree:

I send eunuchs Zheng He and Wang Jinghong with this imperial order to instruct these countries [he stipulated only 'distant lands beyond the seas'] to follow the way of heaven with reverence and to watch over their people so that all might enjoy the good fortune of lasting peace.

The fleet set out with 300 ships in January 1432, taking the now well-mapped route to Java, Palembang, Malacca, Ceylon, Calicut and Hormuz, with some ships dispatched also to Mogadishu, Aden and Mecca. It was inconsequential enough: a last fling for the Xuande Emperor, who doubtless remembered the grand pageants of the voyages launched in his youth by his grandfather. When Zheng He left the port of Changle, it may have been the last time he ever saw China: he probably died at sea on the return trip.

And that, to all intents, also spelt the end of the Ming age of maritime adventure. Wang Jinghong, Zheng He's deputy and like him a Muslim, led the fleet after the admiral's death. But he was killed in a shipwreck off Java in 1434 on the eighth and final journey of the imperial fleet, and he was buried in that country. The next year the Xuande Emperor died; his son Zhu Qizhen was just seven years old, and there followed a power struggle that almost split the empire apart.

At first the battle was between the court eunuchs and the Confucian officials determined to curb their power. They had good reason: the eunuchs had grown wealthy through corruption and embezzlement, and anyone who opposed them was liable to be executed or banished. One of the worst perpetrators was Zhu Qizhen's tutor Wang Zhen, who was as incompetent as he was venal. Through his disastrous advice, the young emperor, who took the name Zhengtong, was captured by Mongols during an ill-advised campaign against them in the north-west in 1449. His younger brother Zhu Qiyu was installed as emperor for the interim, but when Zhu Qizhen's release was secured a year later, he returned to Beijing to find himself placed under arrest in the Forbidden City by the new emperor (Jingtai), who had decided that he rather liked ruling China after all. Yet when his treacherous brother fell ill, Zhu Qizhen seized the opportunity to execute a coup, and he regained the throne that he then held for another seven years.

It was scarcely surprising that, among all this turmoil, seafaring did not seem a priority. These were tough times economically, and a massive flood of the Yellow River in 1448 deepened the problems.

Harassment of the northern borders by the Mongol and the Manchurian barbarians was draining resources. The Zhengtong Emperor favoured the Confucian party at court, who advised that taxes would be better spent developing the nation's agriculture – and moreover, that curbing sea trade would constrict the resources of the eunuchs. The navy was neglected and fell to pieces, and it was forbidden to venture to sea in a ship big enough to make a substantial voyage. In 1500 building a sea junk with more than two masts became a criminal offence, and in 1525 coastal authorities were instructed to destroy all sea ships and arrest any merchants who sailed in them. By 1551 going to sea in any multiple-masted ship was also banned.*

The victors took charge of history, as they always do. According to a late Ming document called the *Kezuo zhuiyu*, in 1477 the Confucians in the imperial court declared that Zheng He was a fantasist and commanded that all the official records of his voyages – full of 'deceitful exaggerations of bizarre things far removed from the testimony of people's eyes and ears' – be burnt. This act of vandalism was allegedly motivated by the efforts of the eunuch Wang Zhi, head of the imperial secret police, to use the records to drum up enthusiasm for a return to seafaring. It's not clear how much faith to put in this retrospective story, but nonetheless there must once have existed plans for the treasure ships that are now lost. If some of the records of the Ming voyages were indeed destroyed, this might have been not so much an attempt to suppress the 'maritime party' as to undermine the influence of the eunuchs who happened to be a part of it.

Zheng He's exploits weren't quickly forgotten, however. A 1597 novel by the popular writer Luo Maodeng called *Xiyang ji* (*Chronicle of the Western Ocean*) retold them as a fantasy filled with magic and derring-do. The eunuch admiral became a cult hero during the late Ming era, when he was celebrated in theatre and literature as the

---

* The anthropologist Jared Diamond attributes to geographical factors not only the different attitude of Europeans to sea travel but the fact that the Chinese voyages were terminated so precipitously. It was a consequence of such an immense unified state, of a sort rendered impossible in Europe by a complicated coastline and obstructive mountain ranges. Unification, he says, has its advantages, but one disadvantage is that a central command can essentially abolish all options. While Christopher Columbus was an Italian entrepreneur who could take his proposals to courts all around Europe until he found sponsors in Spain, there was no equivalent 'market' in China: you were backed by the state or by no one at all.

paradigmatic virtuous commander. But when the Qing historians came to give their version of their predecessors in the *Ming shi* in 1739, they presented the voyages as a great waste of time and money. However impressive the riches that Zheng He brought back, the annals say, 'they did not make up for the wasteful expenditures of the Middle Kingdom'.

With the termination of seafaring, Chinese culture, formerly expansive and enquiring, for a short but crucial period looked inward. While the Western Age of Exploration stimulated curiosity about the world that eroded the rather rigid intellectual inheritance of the Middle Ages and eventually helped usher in the age of early modern science, in China that opportunity was lost. Some have argued that the iron grip in which the imperial court held the country foreclosed the kind of wide-ranging critical enquiry that presaged the emergence of modern science in the West. That overstates the case – the reality was that an emperor's power often rested on a fragile foundation, was prone to factionism, and weakened with distance from the capital. But in any event it is not hard to imagine that a China engaged in a web of maritime trade and intellectual exchange at the same time that the West was discovering the world might have followed a very different trajectory: one that might have brought it into conflict with the expansion of Western interests in South East Asia in the seventeenth and eighteenth centuries, but also into contact with Western ideas. How much the famous 'Needham question' – why modern science developed in the West and not the technologically advanced East – depends on such factors is a debate that shows no sign of reaching a conclusion. But China's turning inward after the high Ming period was clearly not for lack of its ability to negotiate the waters of the world. Whatever the reason for the demise of the Ming voyages, Needham is surely right that the decision 'had far-reaching results not only for Chinese but also for world history'.

But these missions left a legacy, accelerating the spread not only of Chinese culture but also of Chinese people throughout South East Asia. Other countries began to use the Chinese systems of timekeeping, calendars, weights and measures; they began to read Chinese books and play Chinese musical instruments. Some of Zheng He's sailors settled in the places they visited, taking local brides. When the fleets were disbanded, a few officers and crew sailed off with their families in search of better prospects for their trade. Although emigration was officially banned in the early Ming period, nonetheless the fledgling Chinese communities in Vietnam, Siam, Java and the Malay Peninsula grew considerably in the fifteenth century.

Moreover, trade didn't cease despite the ban on maritime travel. The prohibition did not last long, and when it was revoked in 1567 there was a flood of applications for permits from merchants who had been continuing their business illegally in the meantime. The southern Chinese have had an enduring thirst for overseas commerce. After another short period of prohibition during the early Qing dynasty, beginning in 1662, was revoked in 1684, merchant ships were built again in great haste and set off across the South China Sea to Siam, Malacca and the Philippines. Even by 1685 the English pirate and adventurer William Dampier reported that most of the merchant ships in Manila were Chinese, and that many Chinese merchants, shopkeepers and craftsmen had set up residence in the port city. By the eighteenth century the Chinese were doing as much trade in this region as the English or the Dutch were in Europe.

Within the Chinese diaspora, Zheng He remains the maritime equivalent of China's semi-divine water heroes. Even now there is a rather fetching temple to him in Semarang, central Java, called Sam Poo Kong: a derivative of San Bao Dong ('Three Jewels Cave'), since it was allegedly founded by Zheng He (San Bao) himself during a visit. It is used today by worshippers of various religions, including Muslims and Buddhists, and by inhabitants of all ethnicities. It would be naive to suppose that Zheng He himself had such an ecumenical goal when he sailed the seas. But there is nonetheless something fitting about the image.

6 Rise and Fall of the Hydraulic State

Taming the Waters by Bureaucracy

Hydrological systems kept twisting free from the grip of
human would-be mastery, drying out, silting up, flooding over,
or changing their channels . . . No other society reshaped its
hydraulic landscape with such sustained energy as did the
Chinese, nor on such a scale, but the dialectic of long-term
interaction with the environment transformed what had been
a one-time strength into a source of weakness.

Mark Elvin

As dusk deepens to night, the moon's waxing crescent rises above the
rooftops. The constellation that the Chinese call the Jade Rope, a part
of Ursa Minor, hangs low in the sky. Linqing, a retired and ailing
former bureaucrat in the Qing administration, trims the lamp wicks
with a fine pair of scissors and sits down in the Withdrawing to
Contemplate Pavilion in his garden on the outskirts of Beijing. These
days, while the early autumn weather is still warm enough, he likes
to settle here in the evenings and read the *Gazetteer of Famous Mountains*
or the *Classic of the Waterways* – descriptions of landscapes far away,
to which Linqing can now only travel vicariously. But as he unrolls a
scroll, he catches a glimpse of a human form nearby, and turning his
head he sees a young servant boy curled up asleep.

He stands there for a moment, lost in thought, then closes the window against the rising chill and turns up the lamp. The young lad, dozing harmlessly yet in a somewhat impertinent location, has reminded Linqing of the poem 'Admonishing a Son' by Zhuge Liang, the famous military strategist of Shu during the Three Kingdoms period. There was a couplet in that poem which had always bothered him:

> Without tranquillity, there is no way to clarify one's purpose;
> Without peace, there is no way to extend one's knowledge.

How, he had wondered in his youth, could one clarify or extend anything through qualities as passive as tranquillity and peace? Surely action was the key to achievement? But as Linqing records later in his journal, 'Now, after more than thirty years of official service, I finally understand.'

Linqing did not feel particularly peaceful or tranquil in his retirement. Rather, he was disillusioned. Like many state bureaucrats, he had had his share of disappointments, humiliations and failures. As Governor General of the Yellow River–Grand Canal Conservancy, he had lost his job in 1842 when the Yellow River had breached the dykes and poured away northwards, wreaking havoc with the Grand Canal. He had been able to retire with a little grace, however, since he had been recalled to help with repairs to an even worse breach that occurred the following year. All the same, it was clear to Linqing how arbitrarily the heavens played with the fates of government officials, who, no matter how competent they were (and Linqing lacked neither ability nor diligence), couldn't hope to tame China's Sorrow.

In the face of such unpredictability, what good did it do for a man to tell himself that he could effect lasting change through resolute action? 'He no sooner takes action', Linqing wrote,

> than obstructions appear everywhere; before many years [his plans] all come to a halt. But before they stop, no small amount of harm and chaos result. Truly [such a man] does not understand the meaning of the line in the *Great Learning*: 'Only after tranquillity can there come peace.'

The *Great Learning* (*Da xue*) is one of the classic Four Books of Confucianism – like all Qing civil servants Linqing was well versed in the Confucian canon. The quote seems to endorse a Daoist brand of Stoicism, if that chimera may be permitted: a calm acceptance of fate. Yet it is not exactly capitulation to nature's vicissitudes that Linqing had in mind, but something with a more Confucian flavour: a socially gracious renunciation of aggrandizement and fortune. For Linqing believed that the turbulence of his career owed as much to the treacherous currents of state bureaucracy as it did to those of the Yellow River's murky waters. All his life he had sought to do his job well in the face of puffed-up officials, corrupt profiteers and depraved sensualists who indulged their pleasures at the state's expense. 'They [all] lack tranquillity', Linqing wrote, and so 'cannot clarify their purpose'. The perfect Confucian gentleman acts not with striving but with an almost weightless poise and serenity. That is what Zhuge Liang has been telling him. But now it seems too late to profit from the advice.

Linqing is disturbed by a sound: the boy has woken. 'I rolled the scrolls and left the pavilion', he wrote. Soon after that evening, he left this world altogether.

Water conservancy in the Ming

By situating their capital in Beijing, the Ming (1368–1644) and Qing (1644–1912) emperors heightened the urgency of water management, for the canal and river network linking the south to the north had then to be navigable. There was logic to the decision beyond whatever caprice motivated the Ming Yongle Emperor: it was unwise to place too great a distance between the imperial court and the troops protecting the northern frontier, since if the army was isolated then there was always a risk of generals fomenting rebellion. And for the Qing – the Manchu conquerors from the plains north of the Middle Kingdom – the south was an alien territory. Whatever the case, the integrity of the Grand Canal and Yellow River transport system became vital. By the middle of the eighteenth century, more than half of the grain consumed in the capital of Beijing was brought by waterway from the south, which meant that the way had to be kept open at almost any cost. And what a cost it was: maintenance of this transportation channel now accounted for around 10–20% of total government expenditure. 'By binding their strategic well-being to Grand Canal transport and Yellow River control', writes historian Randall Dodgen, 'the rulers of the Ming and Qing dynasties linked the symbolic and the pragmatic to an unprecedented degree. Inevitably, managing the Yellow River became one of the central tasks of imperial administration.'

Dykes to constrain the lower reaches of the Yellow River, attributed to the foresight of Duke Huan of the state of Qi, were built as early as the seventh century BC in the Spring and Autumn period. Two of China's early water heroes, the Han engineers Jia Zhang and Wang Jing in the first century AD, supervised the strengthening of the dykes and construction of sluice gates to draw off irrigation water and reduce the flow. The Song rulers began dredging the Yellow River to try to prevent the bed from rising, although they suffered a massive dyke break in 1194 that changed the river's course. Dredging and reorganization of the channel banks became more systematic under the Yuan in their efforts to build a stable, integrated system of waterways linking the Yangtze, Huai and Yellow rivers.

Removing silt from the Yellow River demanded some impressive technology, not to mention serious organization. The Song government set up a Yellow River Dredging Commission in 1073, which began to deploy boats equipped with dredging tools. The vividly named 'iron dragon-claw silt dispersing machine' was a great rake pulled along the riverbed to agitate the silt and return it to the flow. This principle was extended with the 'river-deepening harrow', a 2.5-metre-long rotating beam fitted with iron spikes, like a thresher for riverine mud. The Ming imperial censor Chen Bangke introduced new techniques in the late sixteenth century, such as wooden machines

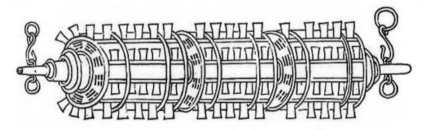

A dredging device used during the Song dynasty. It would be rolled along the riverbed, drawn by cables attached to each end.

set rolling and vibrating by the current to constantly stir up the sediment. In the dry season Chen proposed simply digging out the silt manually.

In 1471 the Ming bureaucracy appointed a Governor General to oversee the Yellow River and Grand Canal conservancy. This new administration considered rival theories of flood prevention. The Han-era Daoist Jia Rang argued that one should manage rivers using the principle of *wu wei* (page 93), giving the waters room to find their natural course. Rivers are like the mouths of children, he argued: try to stop them up and you either make the child yell louder or suffocate it. His philosophy of flood control was to build many irrigation canals that could take up the excess, and to leave the river to carve out a wide floodplain between generously spaced banks. Confucians, in contrast, argued that dykes should be high and narrow, forcing the river to do man's bidding. The tension between these opposed views has never gone away. 'During twenty centuries the two schools contended', Needham wrote, 'and neither proved wholly successful.' But there were other options too, not necessarily exclusive. One could cut diversion channels to ease the flow, perhaps reaching all the way to the sea. One could create reservoirs to hold the excess water in the flood season. One could actively dredge the riverbed to stop it rising. Which way was best?

The Ming official Pan Jixun (1521–95), who was commissioner of the river conservancy on and off from 1565 to 1580, came up with an answer that was heeded for years to come – and which has argu-ably made him China's greatest historical water hero. You don't need to remove the sediment by laborious dredging, he argued: the river can do that for you. If it is confined to a narrow channel, the flow will be fast enough to scour silt from the riverbed of its own accord. The idea was called *shu shui gong sha*, or, as one might say, 'restrict the current to attack the silt'. It had been expounded in the *Zhou li* (*Rites of Zhou*), a classic Confucian text of the Han era describing institutions and governmental principles: 'A good canal is scoured by its own water.' Meanwhile, the *Zhou li* also advised, some of the sedi-ment might be deposited on the dykes to reinforce them. That way, the river is working for you.

But this isn't enough by itself, Pan acknowledged. It's wise also to erect secondary dykes some distance from the river to contain the floodwaters if the main banks are breached. The only problem is that this leaves an expanse of fertile, sediment-enriched land between the primary and secondary barriers, which peasants are apt to start cultivating. Before you know it, what was supposed to be a flood reservoir has become settled farmland: and that is courting disaster. This tension between the needs of flood control and agricultural subsistence has remained an unresolved problem in the river valleys for centuries.

Pan also proposed that Hongze Lake in Jiangsu, into which the Huai River discharged, could serve as a reservoir for storing water that could be released back into the Yellow River to flush its silt. This scheme demanded some remedial work. By 1494, so much of the Yellow River's flow was being diverted into the Huai that the silt expanded Hongze Lake and eventually raised its bed too high for the Huai to flow into it at all. To restore Hongze as a flushing reservoir, it first had to be thoroughly dredged.

Pan's vision was thus not so much a matter of keeping nature in check, but of turning the entire Yellow–Huai system into an almost wholly human-managed plumbing and drainage network. Until the advent of modern construction machinery, the Yellow River Administration adhered to this scheme of active control by means of a massive workforce: a high-maintenance enterprise that demanded constant vigilance and committed rulers to enormous expenditure. Holding back the rivers was equated by one Ming official with holding back the barbarian hordes beyond the country's boundaries: 'building embankments on the Yellow River is like constructing defences on the frontier, and to keep watch on the dyke is like maintaining constant vigilance on the frontier.'

The Confucian strategy of building high confining dykes was adopted in the eighteenth century by the director of the Qing Yellow River Administration Jin Fu, who created a paid workforce to ensure that the job was done well. Until the middle of the twentieth century, many of the dykes were made in the same way as they had been in Song or Ming times: with shovel, wheelbarrow, and back-breaking labour. The barriers were piled up from alluvial soil and packed down to an almost concrete-like solidity. They were typically reinforced with plant fibres. A kind of sorghum called *kaoliang* was especially effective, having an extensive root system that acted like netting to retain the soil. The stalks were tied into bundles called *sao*, which became consolidated as they accumulated silt. It was never a permanent solution: over the course of two or three years the *kaoliang* decays and is consolidated by fresh bundles placed on top until it creates a treacherous foundation, rotten and liable to erosion. This friable assembly was usually supplemented by long sausage-shaped baskets called gabions, woven from willow rods, bamboo or hemp tied with rope or steel and filled with stones. Shifting these elements into place and shoring them up with packed soil demanded the cooperation – often unpaid, sometimes enforced – of thousands of local peasants.

While this sounds like a classic example of feudal oppression, in fact dyke maintenance represents an interesting case of how the Chinese sought to balance public interests and private gains. Throughout the middle and late Ming administrations, officials debated how best to divide responsibilities for this public good between the government, the landowners and the people, in general adopting the policy – reasonable within the social structures of the time – that the beneficiary should

Gabions, used as foundations for building dykes in China since ancient times.

bear the brunt of the cost. Under the corvée labour system, every landowner was obliged to provide a certain number of workers for the task, the contributions determined by how much land they owned and how much revenue they were expected to get from it. It was in the interests of the peasants too, of course, but that didn't mitigate the danger and toil.

These principles were, however, hard to enforce and administer. Not only did many landlords live in towns far from their lands, but wealthy families often built their own private systems of irrigation ditches that interfered with the main hydraulic channels. Foreshadowing the pragmatic economics of the post-Mao reform and opening-up policies, the Ming administrator Zhou Fengming argued that there was nothing wrong in private individuals pursuing profit: they would then maintain the dykes out of the enlightened self-interest that is supposed to sustain a capitalist system.

However it was arranged, this was labour without end: a Sisyphean task. Often it must have seemed to the peasants that they had barely patched up the dykes from one outbreak when they were faced with another. Even if the river was contained during the flood season, the dykes needed continual monitoring and renewal. In the Yellow River

Peasant labourers work to contain the floodwaters of the Yellow River in this seventeenth-century silk painting.

valley a popular saying had it that 'dykes unmanned are like no dykes at all'.

The flood bureaucracy

Historians have sometimes considered a dynasty's ability to restrain the Yellow River to be a bellwether of its vigour. In this view, from the waxing and waning of river conservancy one could gauge whether the state was healthy or moribund, weighed down by corruption and incompetence. Some have even discerned a quasi-periodic rise and fall of authority and empire, as if this were some climatic cycle.* Under the influence of Wittfogel's theory of Asiatic 'hydraulic despotism', the historian Hu Changtu suggested that the late Qing dynasty around the middle of the nineteenth century was heading into the latter phase, saddled with a weak emperor (the Daoguang Emperor, 1820–50) and a monolithic, inert water bureaucracy more concerned to preserve its own existence than to keep the river in check.

This is too neat a story, however, coloured by the old belief (and reinforced by Marxist historical determinism) that emperors who suffered great floods had lost the Mandate of Heaven. It's certainly too simplistic to suggest that a great flood was likely to dethrone an emperor. For one thing, social and political instability could precede massive flooding rather than being a consequence of it: if authority was waning, it was harder to muster the collective resources needed to maintain the dykes.

It's true that a devastating flood could be the final straw that broke the back of a weak administration. But it doesn't follow that the state had to be enfeebled and shambolic for such a catastrophe to occur. The truth was that water conservancy, and the Yellow River system in particular, was too immense a problem for any leader to contain. Maintaining the dykes, paying for repairs, and funding relief when floods occurred drained the imperial coffers, soaking up perhaps a tenth of the total state revenue. One can hardly blame the Daoguang

* There is ancient precedent for this notion of a hydraulic cycle of succession: Mencius asserted that there is a seasonality in politics which demands dynastic change every 500 years or so. In China, power has long been considered to have its own natural rhythms.

Emperor for constantly seeking to cut costs, especially since he had to finance two serious crises in the middle of the century: the colonialist aggression of the British and the internal uprising of the Red Turban sect. And he inherited a parlous treasury to begin with. The illicit trade in opium meant money was leaking out of official circulation, and the state finances had been badly hit by the White Lotus Rebellion of 1794–1804, during the reigns of the preceding Qianlong and Jiaqing emperors. To make matters worse, funds intended for the military suppression of the rebellion were embezzled by the Qianlong Emperor's favourite court eunuch, Heshen. (While that business has been paraded as an example of the high-level corruption allegedly rife during the late Qing era, it was probably rather typical behaviour for any dynastic administration.)

Yet there's no denying that by the early nineteenth century the Yellow River Administration was something of an overpriced juggernaut. Ever since its inception in the Ming era, the extensive bureaucracy of treasurers, military officials, managers and clerical staff all sought, in classic mandarin fashion, placement and advancement for their friends, families and households. It says something about the scale of this administration – or, at least, the budget it commanded – that the tradition of paying officials in grain consumed such a large proportion of what was being transported to the capital that it became expedient to make the payments in silver instead. Some staff bought their posts during fund-raising drives, when the overstretched organization supplemented its revenues by selling ranks and titles: hardly a scheme likely to ensure meritocratic governance. Since many high officials had scaled the ladder that way themselves, they had no motivation to bring such practices to an end.

These mandarins were accused of lavish expenditure and corruption, skimming off public funds into their private fortunes or to pay for visits to brothels. There were all sorts of rumours, some of them circulating in scurrilous pamphlets used as propaganda (or entertainment) in court power struggles. One, titled 'Excesses of the Jiangsu Conservancy River Officials', described the decadent party thrown by one of these gentlemen, at which guests dined on live monkeys' brains and camel's-hump soup prepared in the hump of a live camel.

The administration wasn't by any means wholly degenerate and incompetent, though. Part of the problem was that it wasn't a technocracy.

The officials gained their posts by demonstrating their knowledge of the Confucian classics; they had no reason to be versed in hydraulic engineering per se. In other words, there was a dangerously narrow view of what qualified a man for a government position (although in some ways the appointment of ministers in modern Western democracies is not so different).

If the criteria for obtaining a post were sometimes lax or inappropriate, holding on to it was another matter. In the Ming and Qing governments, failure in the line of duty was generally grounds for dismissal and public disgrace, if not worse. As a result, some genuinely able and honest individuals might sometimes work their way to the top. During the reign of the Daoguang Emperor, Linqing was one of these. He was placed in charge of the Jiangsu water conservancy in 1833, and he attempted to make technical competence a condition of advancement for hydraulic officials.

Another of these capable water engineers was Linqing's counterpart in Henan, Li Yumei, who distinguished himself in several water crises under the Jiaqing Emperor. In a climate that could encourage timid and conservative time-servers, Li was not afraid to innovate. (Neither was he afraid to express utter confidence in his ideas and abilities; fortunately this was generally warranted.) He argued that dykes should be made not of packed soil, which could be eroded rapidly, but of sturdy brick. That was more costly and time-consuming, but Li instituted state-run brick kilns to avoid the extortionate rates of private manufacturers. His measures were sound and effective, and his rectitude made him a paradigm of the virtuous Confucian administrator. In 1877 Li was officially created Li Dawang, the 'Great King Li', to whom temples were dedicated. Even in the dawn of the modern age it was still possible to become a river god.

The trials of the water mandarins

But what a difficult and uncertain road it was towards deification! It's no wonder so few made it. However able these men were, they faced a task that was beyond the capabilities of all the available technology or resources. The Yellow River was too awesome a dragon to tame. A major flood brought disgrace on the engineer or governor in command, no matter how conscientiously they had carried out their duties, and some of the best were removed from their post through no fault of their own. Linqing was one of them.

As soon as he received his first position in the river conservancy in 1825, Linqing threw himself wholeheartedly into the role. While some Confucian officials would have sat back, congratulated themselves on their appointment and proceeded to milk the rewards, Linqing began to gather all the knowledge he could find. 'I lined up several books on river control and wrote out summaries of them', he recorded. He toured work sites, sometimes in searing heat, sometimes in disconsolate rain, asking humble but astute questions of any civil or military officials he encountered. If he found some tool or measuring instrument on site, he'd look for someone who could tell him how to use it. Eventually he acquired an impressive library of writings on hydraulic engineering, and, having pored over them all, wrote his own handbook with the decidedly modern title *An Illustrated Guide to River Engineering Tools*, published in 1837. It was illustrated, he said, because you can gain understanding from images that you can't always get from words (and vice versa).

He took the same approach for *An Illustrated Description of the Yellow River–Grand Canal Confluence Past and Present*, which described how the junction of the Yellow and Huai rivers and the Grand Canal (which at that time all converged near Lake Hongze in Jiangsu province) had been engineered since late Ming times. These were practical books, but Linqing knew how to make them accessible to the Confucian literati who made up his intended audience: like governmental briefings on science and technology today, they aimed to express complicated technical matters in terms that would allow bureaucrats to grasp and use them. Linqing became so trusted by the Daoguang Emperor that he was asked to audit the finances of the Jiangsu conservancy and to root out embezzlement and mismanagement in the funds of his own bailiwick. He did not seek to excuse himself when irregularities came to light. By 1841 he was regarded as a model official.

Little good it did him, however, when the following year the Yellow River once again shrugged off efforts to contain it.

Things had taken a turn for the worse for the Daoguang Emperor since 1839, when the British navy sent its warships up the Yangtze to enforce trading of opium imported from the Indian colonies in the First Opium War (see Chapter 7). In the early autumn of 1841 the treaty that had been tenuously negotiated to end hostilities fell apart, and the conflict was about to resume. Just as the British ships amassed in the Yangtze delta, calamity struck in the Chinese heartland. On 2 August, heavy rainfall brought the waves of the Yellow River over the top of the southern dyke at Zhangjiawan in the Xiangfu area of Henan, north of Kaifeng. Just before he died a year and a half earlier, Li Yumei had singled out this spot as a cause for concern, saying that the riverbank was being badly eroded. He had planned to reinforce it, but after his death those intentions came to nothing.

Li's replacement was an official named Wenchong, who had none of his predecessor's expertise. His appointment illustrates one of the great dilemmas for the Daoguang Emperor, whose fears about the corruption rife in the water administration led him to prefer appointees for their good morality and probity rather than because they knew anything about river management. Wenchong was of the first category; indeed, the fact that he had had no previous contact with water conservancy counted in his favour, because it suggested that he was untainted by its deviant tendencies. Wenchong had met expectations when it came to suppressing corruption, abuses and waste (he even discovered that Li Yumei had been creative with his bookkeeping, albeit simply to ensure that he had funds immediately to hand when needed). But he could not pretend to know much about the upkeep of the river defences.

Floods travel fast. The Xiangfu breakout happened in the evening; by the next morning, the people of Kaifeng found the city besieged by muddy water. As we will see (page 189), Kaifeng was no stranger to floods, and it was surrounded by an earthen dyke some distance from the city walls. But that barrier was not big enough to hold back a serious outbreak of this sort – its purpose was rather to absorb and deflect the impact of the flow so that it would not strike with full force the old, eroded main walls of the city. In consequence, the water ran over the top of the dyke and was soon washing around the walls, where it began to undermine some of the foundations and threaten collapse. The gates could not keep the water out, and the streets and alleys of Kaifeng were deluged.

There had been time for people from the countryside to make their way to the city, where they had better hope of finding food and shelter. This meant that Kaifeng was soon dangerously overcrowded. Others, especially visitors to the city, tried to leave, buying boats if they could afford it, although they did not necessarily possess the skills to handle the craft and there were fatal collisions with half-submerged trees. Those trees were occasionally the only refuge for peasants, who had to clamber into the branches and hope that a rescue boat would arrive before they froze or starved.

Matters deteriorated. The gap at Zhangjiawan was initially quite small, but six days later the river began to rise again and eventually it carved out a one-kilometre breach, discharging the full flow of the Yellow River over the Huaibei Plain. Now Kaifeng became an isolated, waterlogged island. Ritual demanded that Wenchong be stripped of his title: within two weeks of the breach he was dismissed. The same fate befell his superior Bu Jitong, who had been making inspections elsewhere on the night of the outbreak; the same for the governor of Henan, a man named Niu Jian. But the region could hardly cope with the crisis if there was no one in charge, and so the disgraced officials were ordered to stay in their posts anyway and get on with the job – a thankless situation that only set them up for more failure and humiliation.

When a river floods, it doesn't stop flowing even though its full force is dispersed. It doesn't cease eroding either. Freed from its bonds, the Yellow River went right on wearing away the ailing walls of Kaifeng, just as it undermined the reputations of the men chosen to defend the city – and the authority of the emperor whom they served. In early September, the north-west corner of Kaifeng's ramparts began to collapse. However ineffective this barrier now was for keeping out water, the wall was at least absorbing the force of the flow; without it, the waters would sweep unchecked through the city, carrying away buildings. Faced with this impending calamity, the city officials did what desperate people are inclined to do: they seized whatever came to hand to shore up the defences. They reinforced the damaged section with anything they could find: stones and bricks torn from houses and temples or even from the city battlements. Niu Jian convened services of worship to implore the river gods for succour. Maybe his appeals were heard, for the current crashing into the north-west wall soon split into two streams that moved further away, leaving the defences standing – but only just.

With the city under two metres of water, the governor set up food kitchens and sent out boats to fetch supplies from nearby towns. It was a desperate situation. What now?

Wenchong decided that there was no hope of stemming the floods yet, at least not without crippling costs. Instead he recommended that Kaifeng and the environs be evacuated. As for the longer-term question of how to prevent such a disaster recurring, he had a change of heart. The Confucian approach of trying to confine the waters in a complex system of narrow dykes only made the flood worse when it came, he said; what was needed instead was to copy 'the Great Yü's superior strategy': the Daoist principle of 'guiding the stream in accord with the current'.

This was interpreted as pure opportunism on the part of the hapless river official – who, his critics pointed out, had yet even to visit the stricken Kaifeng in person. By proposing to abandon the city and let the river be, Wenchong was accused of trying to make the disaster someone else's problem. For if the floodwaters poured into Hongze Lake, the silt would eventually clog up the crucial section of the Grand Canal linking the Yellow River to the Huai. Hearing of Wenchong's plans, the emperor was enraged. The official was sentenced to stand wearing a heavy wooden collar every day for three months on the riverbank as the autumn hardened into winter, after which time he was exiled to hard labour in China's equivalent of Siberia, the region of Ili in the far north of Xinjiang. Four other officials in the region, accused of negligence or incompetence, met the same harsh fate. It's hard to say how much they deserved it. Not only was it difficult to find any particularly effective strategy in the face of such an overwhelming disaster, or to know how best to direct the limited resources, but officers like Wenchong could hardly be expected to know what decision to make, given that their inexperience in water management was partly what got them appointed in the first place.

To be in the business of water conservancy was, then, not at all the sinecure it was often made out to be. Certainly you might enjoy an official's privileges for a time, but personal disaster loomed every time the summer rains began, no matter how diligently you prepared for it. Certainly there was incompetence, sloth and fraud. But you were in some ways more vulnerable if you were skilled and hardworking, for you would then probably be promoted to the kind of position in which the full burden of responsibility would come crashing down on you when things went wrong. And it becomes harder to condemn a spot of embezzlement or inventive bookkeeping here and there when we recognize that not only were salaries sometimes ill-matched to the real demands of office, but a mishap such as a major flood could see officials reduced to unemployable penury unless they had taken the precaution of storing up a personal financial cushion. Even if officials did not lose their job in such instances, they might be ordered to pay for extravagantly expensive repairs out of their own pockets. Such measures went beyond a simple matter of accountability: they might be an emperor's only means of getting the requisite work done without the necessary funds. In other words, at least some of the blurring of personal and state finances in the water bureaucracy was less a matter of malpractice and more one of pragmatism. Even the emperors understood this to some degree, and they might turn a blind eye to profiteering so long as the job got done.

For such reasons, says Randall Dodgen, we should be sceptical of a caricature that makes the Qing water conservancy a decadent and inept behemoth. Some of those accusations were motivated by vested interests anyway: the waterways of the Yellow River and Grand Canal took potential trade away from maritime transport of grain and goods up the coast. Some charges were self-evidently implausible: it is hard to imagine (as was occasionally suggested) that river officials would allow floods to happen intentionally so that they could skim off profits from the repair work, given that a flood was more likely to bring ruin, disgrace and exile on those in charge. If some officials were conservative, that's natural enough when the alternative of radical change would be more likely to make them seem rash and directly culpable if things went wrong. Why risk innovation rather than simply try to avoid obvious blunders? And if rules and red tape did proliferate during the Ming and Qing eras, at least some of those regulations may have been designed to protect officials from the arbitrary whims and far-reaching powers of the emperors.

One feels, then, for men like Linqing and Li Yumei, who appear to have worked with integrity and a fair amount of intelligent imagination on a task that was all but hopeless. The river officials were perhaps uniquely vulnerable to China's merciless climate and geography: a flood created scapegoats in a way that famines, droughts or plagues of locusts rarely did. Arguably that was the result of the very effectiveness of Chinese hydraulic engineering: it was only because river management began to seem possible at all that inevitable failures were an occasion for blame.

This sword of Damocles compelled some officials to desperate and even fatal extremes. In the Zhangjiawan outbreak, one Xin Decheng, a sub-lieutenant at the Zhonghe prefecture of the Yellow River, was so frantic to salvage his name and career after being stripped of his rank that he supervised the erection of barriers day and night without sleeping, often immersed in the cold waters. When he succumbed to exhaustion and died, he was praised as a great hero. The cost of an official's rehabilitation could be his life.

The 1841 floods were eventually stemmed, but at huge expense in materials, labour and human life. Diversion canals were dug and new dykes were constructed, but as the muddy autumn turned to bitter winter the conditions that the workers must have endured are unthinkable. They needed pickaxes to break up the frozen ground and the ice floes that jammed the streams. Efforts to finally close the breaches in February 1842 were interrupted by storms that battered down some of the repairs and drowned hundreds of workers in freezing water. Not until mid-March was the final plug put in place – a critical operation called 'Closing the Dragon Gate'.

The respite lasted just five months. Then in August it all began again. A storm on the 22nd forced a 600-metre hole in the Yellow River dyke at Taoyuan in northern Jiangsu, which allowed the torrent to pour into the Grand Canal before finding its way to the coast sixty kilometres from the previous outlet.

Linqing was among the officials held responsible. He had been unable to carry out his usual tours of inspection as the flood season approached that summer, because he was called south to perform duties in the conflict with the British forces on the Yangtze during the First Opium War. That was deemed no alibi, and he lost his job and rank, although he did not suffer the indignities of poor Wenchong.

The last battle

Linqing's punishment was, however, soon rescinded. In the following year he was recalled to help with an even more catastrophic outbreak caused by heavy rains and strong winds in July. It happened in Zhongmou, forty kilometres from Kaifeng in Henan province, a region notoriously flood-prone. The Zhongmou flood of 1843 put vast tracts of arable land in Anhui province underneath a sea of silty water two to three metres deep, and famine loomed. No doubt because his competence had never been seriously in doubt, Linqing was pronounced 'capable of self-renewal' and ordered to administer the repairs.

The conditions were almost impossible. Resources – human, financial, material – had been severely depleted by the Zhangjiawan and Taoyuan floods. Despite the desperate situation (indeed, probably because of it), officials and contractors didn't hesitate to cheat and thieve their way to profit. Suppliers of materials, knowing how urgently their wares were needed, cynically put up their costs. These are problems that the massive engineering projects of China have never solved. The flip side was and is that no one really looks out for the labourers. Their pay was indeed often meagre, and they had to cope with terrible weather in little more than flimsy trousers, jackets and sandals. One official described the workforce at Zhongmou in February 1844: 'clothes thin and bellies sunken, forced by hunger to come out [of their mean little tents] in search of food, lying frozen where they had fallen'. They were hired and laid off just as needed, with no thought to their livelihoods. The local population paid the price of such neglect, enduring bands of hungry, unemployed men roaming the countryside. This social unrest was what the leaders feared most from a flood.

The Zhongmou deluge was nowhere near to being checked when winter arrived, and that year saw a bad one. The repairs to the dykes were still not completed by March, when storms caused the waters to rise again, yet already the financial cost had been immense. As the spring floods loomed, it was decided that the repairs would be temporarily suspended, and some repaired dykes were even dismantled to avoid loss of precious materials. But by the time the summer rains were over in August and it seemed possible to return to the job, the Daoguang Emperor had had enough. 'Year after year,' an imperial statement complained, 'military and river repair funds have been needed at the same time.' Couldn't the repairs wait until next year? Alas no, the officials in charge of the project insisted – the job had to be finished now if things were not to get even worse. And so it was, although again the repairs dragged on into the winter, and ice floes claimed more lives.

The 'Dragon Gate' on the Yellow River was finally closed in early February 1845. The officials had their reputations restored – Linqing was among them, although he was returned to a lower rank than previously – and the river was once again confined within its banks. But the Zhongmou flood had damaged more than harvest and home. It had drained the treasury, discredited the river administration and exhausted the emperor. 'Although a gloss of normalcy was restored in the years after 1845,' writes Dodgen, 'the Yellow River control system and the bureaucracy that maintained it were living on borrowed time.' So too was the Daoguang Emperor himself – he succumbed, aged sixty-seven, in 1850. He is generally remembered now as a well-meaning but ineffective ruler who never came to grips with the realities of the nineteenth century.

What a curious, even desperate situation this now appears. Here was a labour force using methods little different from those of the Han dynasty, overseen by a creaking bureaucracy that would have been rich material for the satire of a Chinese Dickens, and which still considered it an essential precaution to cast offerings of rice cakes into the water the placate the river gods. All this in a country smarting from a humiliating defeat under the modern firepower of Western barbarians. The modernizers of China's twentieth century surely had reason to criticize their country for being mired in the past. If it is no wonder that China in the late Qing era looked as exotic and strange to Westerners as did the Yuan empire to Marco Polo, they could have comprehended the nation more clearly had they appreciated that its customs and institutions had been mobilized in large part to deal with the unmanageable excesses of its hydrological environment.

At any event, the 1843 floods presaged the collapse of the water bureaucracy. They had crushed the Daoguang Emperor, and his son and successor, the Xianfeng Emperor, showed no appetite to return to the fray. After further flooding in the early 1850s, the Grand Canal was largely abandoned for transport of tribute grain, which was carried instead on the maritime route up the coast. After another great flood on the Yellow River at Tongwaxiang in 1855, no attempt was made to repair the dykes – instead the river was permitted to remain in its new course, some 800 kilometres north of its former channel, which it still occupies today. Instead of building high dykes as flood barriers, only low ones were erected to dissipate the force of future breakouts.

Let the river build its own banks of silt, the head of the Henan conservancy argued – the deposits will create new farmland too (albeit in constant danger of inundation). It was less a triumph of Daoist over Confucian engineering, more a tired admission of defeat. River management was all but abandoned, and the Yellow River Administration was abolished in 1861, leaving local authorities to cope as best they could. The river, it seemed, had won.

Flood brothers

The dissolution of Yellow River management in the latter half of the nineteenth century has been read as a metaphor for the fate of the Qing empire itself. After the glories of the Kangxi, Yongzheng and Qianlong emperors of the eighteenth century – the three 'great Qings' – there was nothing but slow decline, of which the debacle of the Opium Wars was just the opening act. This is a fair characterization in broad terms, but it's hard to say if the image points to any cause-and-effect relationship. Did a fading dynasty lose the power and authority, indeed the mandate, needed to manage the rivers? Or did the unruly waters themselves fatally weaken the emperors? Probably a bit of both: it's too simple to suggest that the Qing rulers lost control of the rivers through apathy and weakness, but once they did, their position became ever more precarious.

The Xianfeng Emperor inherited a profoundly unappealing throne in 1850. Not only were the hydraulic works in total disarray and the state beholden to foreign colonial powers, but the Taiping Rebellion was just erupting. It was to become the worst civil war in history, claiming an estimated 20–30 million lives – more than the First World War. The Taipings were considered by the Qing rulers to pose a far greater threat than the British, who were determined only to subdue them and not to supplant them. 'The British are merely a threat to our limbs', said the Xianfeng Emperor's brother, but 'the rebels menace our heart'.

Famine caused by neglect of the irrigation systems seems to have played a part in sowing the discord on which the Taipings drew. But there were other dominoes ready to fall. British opium wreaked havoc with traditional social structures, and the Chinese defeat had flooded the marketplace with imported commodities, causing local industries to decline and shed labourers. All these problems were, rightly or wrongly, laid at the door of the Manchurian rulers.

Whatever the composition of the mixture, no one could have predicted the nature of its turbulence. Arguments in southern China for deposing the Qing overlords weren't hard to find, but the one offered by the Taiping leader Hong Xiuquan scarcely seemed the most persuasive. Hong insisted that a foreign god – the deity of the Christian missionaries – had revealed that he, Hong, was the brother of Jesus and had told him to eliminate the Manchurian demons. The Taiping uprising was essentially a millenarian movement of the kind that had convulsed Western Europe in the Reformation: utopian, populist, mystical and deeply religious. Hong promised a Heavenly Kingdom of Great Peace where all would be equal and private property was forbidden.

Hong encountered Christianity via missionary teachings while visiting Guangzhou in 1836 to sit (and fail) the civil service examinations. The message of Christ was still a novelty in China at that time, a 1724 ban on missionaries by the Yongzheng Emperor having only recently been lifted. And what could be more propitious than that Hong's family name (洪) means 'flood'?* Hong came by a Christian tract distributed by a missionary in which, in Chinese, he read the story of the Flood in Genesis, which told him that a *hong* had destroyed and renewed the world. More, Hong's given first name, Huoxiu (he renamed himself Xiuquan after his prophetic, delirious near-death visions the following year), contained the same character (*huo*, 火, meaning water's opposite, fire) as the name of the God described in this text: Ye Huo Hua, Jehovah. As the First Opium War was about to ignite, Hong saw the events in apocalyptic terms, deciding that God had chosen him to overthrow the Manchu tyranny. When the Christian sect that he formed in 1843 began to suffer persecution from the authorities, it burst into violent rebellion.

Unlikely though it might seem that the brother of Christ should turn out to be an ethnic Chinese peasant from Guangdong, in other respects it is perhaps surprising that there were not more of these

---

* It may also mean simply 'vast': that a flood should be equated with that concept speaks for itself. It appears in this context in the name of the first Ming Emperor, Hongwu – which explains part of its significance to Ming loyalists in the Qing era.

communitarian revolts within imperial China. Anti-Manchu feeling and Han nationalism were widespread, particularly in the south; Guangdong had also brewed up Ming loyalist sects such as the Red Turbans and the Heaven and Earth Society. The messianic preaching of Hong Xiuquan and his companion Feng Yunshan resonated particularly with the latter – not only because, like those two men, the Heaven and Earth Society was drawn largely from the Hakka people of Guangdong and its environs, but from a further linguistic coincidence. The society, like the Green Gang bandit brotherhood of Jiangsu, used a secret code word: *hong* (洪). They were in fact often known as the Hongmen (*men* here meaning gate or portal), by which name they continue to exist even today in Taiwan (where they constitute a Masonic-style political fraternity) and Hong Kong (where they are illegal because of their association with the criminal underworld organization the Triads).

To emphasize their rejection of the Qing, the Taipings refused visible trappings of Manchu dominance such as the hair worn in a pigtail. They introduced some commendable measures, such as outlawing opium and improving equality between men and women. Hong also banned the ancient practice of foot-binding.

Like uprisings of the past, Hong's revolutionaries recognized the importance of controlling the waterways. They marched north into Hunan and attacked Changsha on the Xiang River, building pontoon bridges so that the besieging forces could come and go on both banks. Although the Qing forces held the city, Hong assembled a large navy from the ships and boats captured from Changsha's wharves and moved on up towards Dongting Lake and the Yangtze, mastering the art of patrolling the rivers to guard against ambush. As they approached Wuchang, the capital of Hubei on the Long River, they evolved a formidable, fluid skill in river warfare. Historian Jonathan Spence describes it as consisting of

> taking to the land when least expected, abandoning fleets of hundreds of boats at one spot only to seize new fleets a thousand strong when they descend upon some unsuspecting river town, cutting bridges as they pass them to delay pursuit, recruiting the boatmen along with the boats to check the Qing, throwing up pontoon bridges where no other bridges had existed, then removing them and moving and floating them downstream to use again.

Once the Taipings seized the middle Yangtze by occupying Wuchang and Hankou, they began to look unstoppable. In February 1853 they set out from Wuchang with a fleet of thousands of boats, heading downriver towards the still greater prizes of Nanjing and the Grand Canal junction at Zhenjiang and Yangzhou. By the end of March Nanjing had fallen (Yangzhou followed in April), and it was here that Hong set up the Heavenly Capital of his Earthly Paradise, from where he dominated the economic region of the Yangtze basin for a decade.

The Taipings found less success in the north, where they set out to capture the Qing capital of Beijing. Although unprepared for the harsh northern winter, they managed to cross the Yellow River and by October 1853 one column reached the outskirts of Tianjin, only 110 kilometres from the capital. But on open terrain they couldn't best the Qing's Mongolian cavalry, and they were driven back to Lianzhen on the Grand Canal in Shandong. Here the Qing commander ordered construction of a ditch to carry water from the canal into a dried-up river that flowed past the Taiping encampment, gradually making the place waterlogged, then flooded outright. The rebel troops were forced onto roofs and rafts, where they became easy targets for Qing firepower. By March 1855 the northern campaign ended in disaster for the Taipings.

For some time the foreign powers regarded the conflict with neutrality, despite pleas from the Qing to intervene: the British decided that any attempted interference would only harm the prospects for relative stability, and thus for their commercial interests. A tentative expedition of the British naval steamship *Hermes* from Shanghai to rebel-occupied Nanjing initially met with a favourable reception. All the same, when its commander, the British official Sir George Bonham, was delivered a message from the Taiping 'East King' Yang Xiuqing implying with gratification that their fellow Christians the British now 'acknowledge our sovereignty', Bonham replied with a curt rejoinder that British subjects were subordinate to no other sovereign and that any threat or injury to British persons and property would be met with a response like that given to the Qing ten years earlier. He then beat a hasty retreat to Shanghai, where he reported to the Crown that the Taipings' alleged Christianity was 'a tissue of superstition and nonsense'. A French expedition to Nanjing in December 1853 found the reception equally baffling and imperious. An American contingent didn't reach Nanjing until the spring of 1854, where they were told that if they revered heaven and recognized 'the Sovereign' then they would be permitted to bring tribute to the court annually and 'bask in its grace' – and that they should 'tremblingly obey'. On receiving the missive, the US naval captain Frank Buchanan found it 'peculiar and unintelligible'.

Yet a delicate truce with the Westerners persisted, and when the English explorer Thomas Blakiston passed through Nanjing in 1861 he received courteous treatment. Struck both by the Taipings' 'tremendous heads of hair', in contrast to the Qing pigtails, and by the 'gaudy colours of the dresses of both men and women', he was nevertheless sceptical about their prospects. 'I see no hope of the Taipings becoming the dominant power in China,' he wrote, 'because they are simply unable to govern themselves, except by a species of most objectionable terrorism.' Yet neither did he think that the Manchus would ever fully recover their authority. 'Things are governed in China by rules that we don't understand', he concluded.

Like so many millenarian leaders before him, having established his kingdom Hong allowed his utopian ideals to descend into aggrandizement, despotism and libertine excess. He banned from general use the character for his family name, as well as those for 'sun' and 'moon' (others with the same pronunciation were substituted); only Hong himself was the true Sun, and his wife the Moon. Among the other freshly minted characters was one for 'rainbow' (also pronounced *hong*), which depicted a raincloud above the sacred flood-*hong* (洪), reminding of the judgement of heaven – and identifying Hong Xiuquan with both God's wrath and his mercy. Hong himself retold the biblical story in his 'Imperially Written Tale of a Thousand Words', the primer from which children in the Earthly Paradise would be educated:

> Far off, the clouds gathered
> And rain fell from the void.
> After the flood waters ebbed away
> God in his compassion made a Covenant:
> Never again to send such a deluge –
> The rainbow [*hong*] would stand as his sign.

In 1856 Hong moved against his rival Yang Xiuqing, who as Jesus's 'fourth brother' had been steadily arrogating power in the name of their divine Father. Yang was beheaded by the Taiping 'North King', Wei Changhui, who then proceeded to slaughter several thousand of Yang's household and supporters. Yang had attempted to curb the excesses of Hong's rule, in which concubines and other attendants were granted arbitrary powers. Freed from such constraints, he did as he pleased, indulging his megalomania and paranoia. He became obsessed with cleanliness and the threat of disease: he was constantly fanned by servants to keep insects away, though no fan should ever come closer than five inches from his body. Every morning female attendants chanted Hong's poems, recited from memory. Mistakes of etiquette were punished with beatings, which the offender was expected to receive with cheerful gratitude. In short, Hong became as monstrous a dictator as China had ever seen.

The Qing were still fighting in south-central China, and by late 1859 they had laid siege to Nanjing. In an effort to divert the assault, the Taiping commander Li Xiucheng led a naval force to attack Hangzhou, then swiftly returned to Nanjing to fall on the Qing forces depleted by the dispatch of relief troops to Hangzhou. The Taipings then captured Suzhou, and in June 1860 they stood poised to assault Shanghai. Li assumed that the Western powers would remain neutral, especially after assurances on his part that they would not be attacked. He misread their protectiveness of commercial interests: the foreigners opened fire as the Taiping troops advanced, and Li's plans came to rain.

In early 1862 Li returned with a less placatory message: if the British or French tried to prevent his army from entering Shanghai, they would be beheaded. But heaven delivered judgement: in January there was an unusually heavy snowfall and the river froze over, trapping the rebels and rendering any attack impossible. For peasants in the surrounding countryside, many without food or shelter, the situation was appalling.

After the spring thaw, the Taiping army was forced to retreat back upriver to counter a Qing attack on Nanjing. In 1863 the British, led by Charles Gordon, and the French followed in pursuit with gunships and cannons. In an effort to divert the Qing siege, Li Xiucheng launched a desperate campaign to the west in Anhui, but was threatened on the north bank of the Yangtze, now in full summer flow, by a Qing river force equipped with Western armaments. As the Taipings tried to cross the river, starved, exhausted and in disarray, the Qing gunboats opened fire on the densely packed troops massed on the riverbanks and mowed them down without mercy.

Nanjing fell in a bloody assault in 1864, and Hong was dead before his defences crumbled – some say through suicide, although sickness brought on by lack of food seems more likely. He left behind a deeply traumatized nation. Lord Elgin's secretary, Laurence Oliphant, was appalled by the destruction he witnessed on Jinshan Island, south-west of Shanghai:

> The devastation is now widespread and complete. A few peasantry have crawled back to the desolate spots which they recognize as the sites of their former homes, and, selecting the heaps of rubbish which still belong to them, have commenced to construct out of them wretched abodes . . . the destitute appearance of the scanty population served rather to increase than diminish the effect which this abomination of desolation was calculated to produce.

At Zhenjiang, the scenes were even worse:

> Deserted streets, between roofless houses, and walls overgrown with rank, tangled reeds; heaps of rubbish blocked up the thoroughfares, but they obstructed nobody. There was something oppressive in the universal stillness; and we almost felt refreshed by a foul odour which greeted our nostrils, and warned us that we had approached an inhabited street.

Of a former population of around half a million, Oliphant reckoned that barely 500 still lived in the city.

Nanjing itself never quite recovered from the Taiping occupation. Not until 1876 was it restored as a treaty port, during which time Shanghai – which a representative of the East India Company identified in 1832 as having the potential to become 'the principal emporium of Eastern Asia' – usurped its role as the major seaport of southern China. Shanghai never looked back: by the 1930s half of all foreign investment in China was situated there, and it was home to half of all the foreigners resident in the country.

After the Taiping Rebellion was put down, China looked exhausted – and ripe for colonial exploitation. Other nations began to help themselves almost as freely as they did in Africa. The British and Americans negotiated free passage along the Yangtze, while the French obtained concessions in Yunnan. In the First Sino-Japanese War of 1894–5 the Japanese took Taiwan and the Liaodong Peninsula in Manchuria, but the Western powers – Russia, France, Germany – moved at once to lay their own claims in the Triple Intervention of 1895. This gave Russia Liaodong, while the Germans moved into Shandong, taking over the port of Qingdao (where they established the brewery that still makes China's finest beer).

Resentment at the foreign occupation – and civil unrest prompted by drought followed by flood – spawned the Boxer Rebellion in Shandong in 1899. The formidable Empress Dowager Cixi, de facto ruler of the Qing court after she installed her four-year-old nephew as the puppet emperor Guangxu in 1875,* saw this as a chance to throw out the Western imperialists, and she announced her support for the Boxers, who began as another archetypal Chinese secret society. When the European, American and Russian troops suppressed the rebellion in 1900, Cixi fled Beijing, only to be permitted back by the international forces in the hope that the chastened Qing rulers would prevent China from disintegrating.

It was too late for that, and the international resolution of the Boxer Rebellion only hastened the end of the Qing. To pay for the massive reparations demanded by the foreign powers, in 1911 the Qing government nationalized the railways leading west and south from the Yangtze hub of Hankou, precipitating protests and strikes. When a bomb built by a revolutionary group was accidentally detonated in the Russian concession of Hankou in October, these rebels were forced to play their hand, and they staged a coup. Once they were joined by the Qing army stationed in nearby Wuchang, the mutiny gained momentum – and with command of the vital Hankou–Wuchang–Hanyang juncture of the Yangtze, it was invulnerable to the Qing loyalist forces. The Qing rulers were forced to cede power by the end of the year, and the last emperor, Guangxu's five-

---

* All pretence of regency was dropped in 1898 when the Empress Dowager, disturbed by Guangxu's reform plans, removed him from power and in effect placed him under house arrest, even though he remained emperor in title until his early death in 1908.

year-old nephew Puyi, stepped down. The rebel leader Sun Yat-sen, a Guangdong peasant who had been in exile in the USA seeking funds for his anti-Qing movement, returned to assume presidency of the newly formed Republic of China as head of the National People's Party or Kuomintang (Guomindang).

Broken land

Amongst all the turmoil in the second half of the nineteenth century, there was not the slightest prospect that the abandonment of central river management under the Daoguang Emperor would be reversed. Flood control on the Yellow River was not neglected entirely, but it was now focused only on the economically important regions near the coast, and the poorer regions inland were left to nature's mercy. Without adequate maintenance, the utterly artificial nature of the ecosystem on the North China Plain was laid bare: it collapsed rapidly, and the consequences were disastrous. It was during this period that 'China's Sorrow' became a commonplace sobriquet for the Yellow River.

Life in the putative 'cradle of Chinese civilization' was grim. When the British explorer Ney Elias visited the district near the 1855 breach in the river defences, he found 'many entire villages half-buried in deposits and deserted by the greater portion of the inhabitants; those who remain being in a poor and miserable condition'. The houses, he wrote, are 'frequently silted up to the eaves'. With the collapse of water conservancy, drought became as great a danger as flood. The north China famine of 1876–9 was the most deadly in history: around 10–13 million people lost their lives in conditions of unimaginable destitution. They ate roots, bark, ground-up stones – and, not for the first or last time, the bodies of the dead.

These crises appeared to expose China as a nation out of step with the modern age: backward, unable to feed its population, constantly on the edge of calamity, threatened from within and without and struggling for its very existence. A great flood in 1886–7 near Zhengzhou in Henan killed somewhere between 1 million and 2.5 million people, but little was said of it in the Chinese press for fear that it would confirm the international sense that the country was a lost cause. The American newspapers reported that, as the floodwaters eventually found their way into the Huai and Yangtze catchments, they left behind vast alluvial plains almost unbroken in their monotony save for the occasional roof or treetop poking above the silt. After a summer drought that had destroyed the wheat crop, now the millet, beans and potatoes of the North China Plain were wiped out too; the consequent famine sent the death toll soaring still higher.

The problems went beyond a crisis of authority: they challenged the very notion of national identity. When Sun Yat-sen established the Republic in 1911–12, it was clear that China needed an active programme of state-building, which would depend both on taming the waters and on manipulating hydraulic legend into a compelling new narrative.

7 War on the Waters

Rivers and Lakes as Sites and Instruments of Conflict

I observe that water can be used to encroach and inundate, can be used to float and flow, can be used to sink and drown, can be used to encircle and besiege, and can be used to quench thirst . . . The stupid must employ boats and vessels before they term it aquatic warfare, not knowing that if they fathomed its real meaning, prepared their implements, and took advantage of opportunities, aggressive warfare and unorthodox plans would all come out from it. What reliance must there be upon boats and vessels?

Ye Mengxiong, *Yun chou gang mu*
(*Manual of Transport Management*) (*c.*1562)

Sandwiched between the mighty Ming and Qing empires is perhaps the most ephemeral, obscure and equivocal of all Chinese dynasties. For the single 'emperor' of the so-called Shun dynasty, a rebel commander named Li Zicheng, never meaningfully ruled China at all.

In the mid-seventeenth century a series of drought-induced crop failures and an outbreak of plague stirred up social unrest, and disenchanted peasants began to band together to contest Ming rule. These rebels coalesced into two great armies, and Li Zicheng led one of them. Even as he besieged and captured the ancient capital of Xi'an, he maintained the pretence that he was a loyal subject trying to liberate the Ming Chongzhen Emperor from the malign influence of officials.

That fiction became harder to sustain once Li took Luoyang and then Kaifeng, at which point his victory seemed assured.

Now styling himself the 'Prince of Shun', in 1644 Li advanced on the capital of Beijing. Recognizing that defeat was inevitable, the Chongzhen Emperor committed suicide and the glorious Ming dynasty was brought to a close. Yet Li occupied Beijing for barely a month before his army was defeated by the former Ming general Wu Sangui and his allies, the Qing from Manchuria. The rebel chief only just had time to proclaim himself the first Shun Emperor before fleeing to the west and disappearing from history, presumed dead by 1645. A humble peasant of Sha'anxi who proved to be a skilful military strategist, Li learnt that it is easier to end a dynasty than to found one.

Li's campaign saw one of the most devastating uses of water as a military weapon. In 1642 his forces surrounded Kaifeng (then known as Bianjing) and laid siege for many months. Li tried everything. He built a great tower higher than the city wall, armed with cannons, but his opponents responded by building an even taller one overnight to return the fire. He tried tunnelling through the 35-metre-thick walls, but was repelled. He filled the excavations with gunpowder to blast down the walls, but the explosions blew outwards, killing his troops as they rushed forward in anticipation of a breach.

Although these assaults were repulsed, the Ming governor of Kaifeng was getting desperate, and in the summer of 1642 he issued a fateful command. The dykes of the Yellow River, which was swollen and raging in the flood season, were to be broken down so that the deluge would disperse the rebel troops. Having run out of other ideas, Li had already hatched the same scheme: he planned to flood Kaifeng to end the resistance. Neither side seemed to consider that the flood-waters would harm them too, believing that only their opponents would be damaged.

Kaifeng had been the capital of the Song dynasty, and its proximity to the Yellow River had made it a major centre of commerce; in the eleventh century it may have been the largest city in the world. But the Mongol invaders had besieged it and destroyed the hydraulic network that sustained it, and then the Yellow River itself had shifted course, leaving Kaifeng stranded and marginal on the floodplain. Repeated flooding had gradually raised the level of the surrounding land above that inside the city walls, so that Kaifeng was a basin ready to be filled if the river broke its banks.

It was the citizens of Kaifeng who came off worse from the governor's plans to use the river as a weapon against Li Zicheng. The city was drowned to its rooftops. The waters rampaged through the walls and into the streets, destroying homes and sweeping people to their death. The death toll seems hardly credible: allegedly, around 300,000 of the 378,000 inhabitants of Kaifeng perished in this human-made catastrophe. The once great city was reduced to ruins, making Li's victory a hollow one. Devastating famine and pestilence followed the 1642 flood, which has been ranked as the seventh greatest 'natural' disaster in history. Kaifeng was abandoned until it was rebuilt by the Qing Emperor twenty years later, and it never recovered its former glory.

In war, water is a dangerous and unreliable ally. That has rarely deterred Chinese leaders from thinking that they can command its power: a belief all too often proved delusive.

The famous sixth-century-BC martial strategist Sun Tzu (Sunzi), whose treatise *Sunzi bingfa* (*The Art of War**) was allegedly an influence on leaders ranging from Mao Zedong to Norman Schwarzkopf, regarded water's military significance to be primarily metaphorical. 'Military tactics', he wrote,

> are like unto water; for water in its natural course runs away from high places and hastens downwards. So in war, the way is to avoid what is strong and to strike at what is weak.
>
> Water shapes its course according to the nature of the ground over which it flows; the soldier works out his victory in relation to the foe whom he is facing.
>
> Therefore, just as water retains no constant shape, so in warfare there are no constant conditions.

Sun Tzu was rather wary of real water, however ('After crossing a river, you should get far away from it', he advised), and the image of warfare generally presented in *The Art of War* – great armies manoeuvring

---

* Like other texts from the Spring and Autumn period, the authorship is attributed by tradition rather than any real historical evidence. According to Sima Qian, Sun Tzu served in the state of Wu, but whether he was the original author – indeed, whether he was a real figure at all – is uncertain.

over open tracts of land – gives a very incomplete view of how military affairs were conducted in China. Very often the key conflicts involved rivers, lakes and marshes. They were about dominating the water routes, and often took place on the water itself. Great fleets clashed on lakes and rivers in naval engagements comparable in size and significance to any taking place in the seas of Western Europe, or later in the open Atlantic and Pacific. China was a kingdom contested on water, with water, and for water.

I will confess from the outset that this chapter has to relinquish even the rather loose chronology that I have attempted to impose hitherto. It is not a totally feeble excuse to say that this is part of the point. For the issues haven't changed through the ages; the strategic importance of waterways was as great for the Qin conquering Wei and Shu as it was for the Communists and Nationalists fighting the Japanese. (Only with the coming of the railways was their role for military transportation rivalled.) The decisive power of the Tang 'tower ships' and Song paddleboats was not so different from that of the imperial British gunships conquering the Yangtze in the nineteenth century. And attempting to harness water itself in battle was no less hazardous for Chiang Kai-shek in the 1930s than it was for Li Zicheng and the Ming in the 1640s. Aquatic warfare has been a constant determinant of China's fate.

Fire on water

According to Niccolò Machiavelli, 'The main foundations of every state . . . are good laws and good arms . . . you cannot have good laws without good arms, and where there are good arms, good laws inevitably follow.' Few nations were as ready to face up to the realities of power and governance as Machiavelli would have had them do, which was of course why his unflinching view of political power won him many enemies. China is no exception. The Confucian political philosophy stressed that stability depended on the virtue of the emperor; if he was virtuous, 'good laws' would follow and the population would be content without coercion. But in fact the state was often created and maintained by military force: by organized violence and war. With a nation this vast, this vulnerable to uprising, rebellion and invasion, it is hard to see how it could have been otherwise. Time and again, discord and disobedience began far from the centres of power – if an emperor let down his guard or let himself be distracted in one quarter of the empire, trouble brewed in another. While some dynasties did fall through foreign invasion, such as the incursions of the Jurchens (Jin), Mongols (Yuan) and Manchurians (Qing), others failed because of poor leadership and bad policy decisions: the collapse came from within. That is really what the Communist Party of China fears today.

Leaders from the Han to the Maoist eras could affect an attitude of *wu wei*, of remote benevolence, only if in reality they possessed a formidable apparatus of state control. And to control China, you must control its rivers – whether that means recognizing their strategic military and economic importance, or literally restraining them as if nature itself were the enemy.

Waterways served two primary strategic functions. The first was as transportation conduits. The first Qin Emperor, Shi Huangdi, could not have contemplated conquest of Sichuan without relying on the Min, Yangtze and Han rivers to take his troops deep into the kingdom of Shu. The Han and Yangtze were also essential to the Qin campaign against the kingdom of Chu downstream to the east, prompting the Qin general Bai Qi to create amphibious military units around 280 BC. Military goals initially motivated the great Dujiangyan hydraulic waterworks masterminded by Li Bing (see page 109): the crops that it irrigated were needed to feed the troops in Sichuan.

The second strategic role of rivers was as natural barriers to conquest. When the Southern Song rulers in Hangzhou called the Yangtze its 'Great Wall', they were alluding to the obstacle it – and the other rivers in the perpetually contested region between the Yangtze and the Huai – presented to the equestrian forces of invaders such as the Jurchens and Mongols, who were all but invincible on the grassy plains of the north. As the Jurchens descended to create the Great Jin dynasty, pushing the Song into southern China in 1127, it was the Huai River that marked the boundary between them.*

---

* The massive flood of the Yellow River in 1194, which sent it on a new southerly route that eventually entered the Huai River and emptied into the sea south of the Shandong Peninsula, was therefore not unwelcome to the Southern Song, as it restored the lower part of the river into their hands.

For these reasons, China's wars were often waged on and around the rivers. They were the arteries of military conquest, the fluid arenas of dynastic change. Battles fought on rivers and lakes became the stuff of legend. The most famous of them is surely the Battle of the Red Cliff, the engagement in AD 208 that sealed the dissolution of the Han and marks the beginning of the Three Kingdoms period. This was possibly the largest naval conflict in history in terms of numbers of vessels, although the battle has been so romanticized – most famously in the early Ming classic* *Romance of the Three Kingdoms* – that it is hard to distinguish fact from fantasy. Poets from Li Bai to Su Shi have composed odes in its memory; it is, in the historian Lyman van Slyke's estimation, China's Trojan War or Arthurian legend, a pageant of drama, pathos, comedy, loyalty and deceit. Everyone in the Yangtze valley will tell you these stories, which they know better than their own recent history.†

The demise of the Han dynasty was messy. Like many other dynastic declines, it began as a peasant revolt incited by dissatisfaction at the oppressive conduct of a corrupt ruling class, and was exacerbated by flood and famine – in this case caused by breaches of the lower Yellow River. Those latter events were interpreted as a withdrawal of heaven's mandate, and from around AD 170 peasants displaced from their homes by floodwaters and penury, along with unemployed soldiers, formed into bands that swelled to ramshackle armies. In 184 a Daoist rebel sect called the Yellow Turbans began to wrest territories north of the Yellow River from the command of the emperor Lingdi.

The Yellow Turban uprising lasted for twenty years, and by the end of it the Han empire had been brought to its knees. After Lingdi

---

* Regarded as one of the great works of Chinese literature, the *Romance* was traditionally attributed to Luo Guanzhong (*c*.1330–1400), but it seems to have been compiled by several writers up to the early sixteenth century.

† The 2008 movie *Red Cliff* directed by John Woo, the most expensive ever made in China to date, was understandably more concerned to secure the legend than to clarify the history, but the huge cast of characters was as familiar to Chinese audiences as Robin Hood's Merry Men are in the anglophone world. There were allegedly protests when Woo initially cast a Japanese actor for the villain Cao Cao, on the grounds that so important a historical figure should never be played by a Japanese, however wicked he might (traditionally) be.

The precise location of the Red Cliff is not known. It is often considered to have been in Huang prefecture in Hubei, but that was doubted even in the Song era. The debate didn't stop artists from painting dramatic views of the precipitous cliffs, like this one by an unknown artist of the Yuan dynasty around the fourteenth century.

died in 189, rule was shared between his consort Empress He and her half-brother He Jin, general of the Han army. But He Jin was hostile to the powerful clique of court eunuchs, and later that same year he was assassinated. A warlord named Dong Zhuo then seized the throne, ruling through the puppet emperor Xiandi, Lingdi's son. When his harsh and despotic rule ended with his death in 192, another ambitious warlord – Cao Cao of Wei, who had acted as a Han military commander during the Yellow Turban revolt – made Xiandi his own puppet and effectively ran what remained of the empire.

Cao Cao's authority was challenged by the leaders of other states: by Sun Quan, Marquis of Eastern Wu, south of the Yangtze in modern Zhejiang, and by Liu Bei, a warlord who set himself up as ruler of the state of Shu.* Faced with Cao Cao's overwhelming forces, Sun and Liu agreed to an alliance, and they met Cao Cao's troops at Chibi (Red Cliff) on the Yangtze in Hubei. Some records claim that Cao Cao had over 800,000 men, his opponents just 30,000. The outcome of the battle would decide the future of China: would it be unified by Cao Cao, masquerading as a servant of the hapless Xiandi, or splinter into rival states?

---

* This is generally called Shu Han to distinguish it from the ancient kingdom in the same region of modern-day Sichuan and Chongqing.

余懷望美人兮天一方客有
吹洞簫者倚歌而和之其
聲嗚嗚然如怨如慕如
泣如訴餘音嫋嫋不絕如

Many poets have rhapsodised the Battle of the Red Cliff. This is the Song writer Su Shi's 'Ode to the Red Cliff'. Like many such examples, it was intended as a reflection on personal and political travails in the author's own time, expressed through the distancing language of familiar legend.

As the Shu and Wu forces confronted Cao Cao's massively superior army, the Wu commander Zhou Yu played an old trick. To plant an alleged defector in the enemy midst to lead them astray – compare the ploy of the king of Qin against ancient Shu (see page 109) – seems to assume an optimistic degree of credulity. But it perhaps speaks of the fissiparous nature of warlord-era China that such defections were common enough to make the scheme believable. In any event, Zhou Yu sent his military strategist Pang Tong to join Cao Cao. When Pang Tong heard that Cao Cao's army, unused to river combat, was becoming seasick on the ships,* he proposed that the vessels be chained and bolted together to stop them from rolling with the waves. 'The river is wide, and the tides ebb and flow', he says to Cao Cao in the *Romance of the Three Kingdoms*:

> The winds and waves are never at rest. Your troops from the north are unused to ships, and the motion makes them ill. If your ships, large and small, were classed and divided into thirties, or fifties, and joined up stem to stem by iron chains and boards spread across them, to say nothing of soldiers being able to pass from one to the next, even horses

* Cao Cao is said to have had an artificial lake constructed to train his troops in river combat. But apparently this measure was insufficient.

could move about on them. If this were done, then there would be no fear of the wind and the waves and the rising and falling tides.

Then Pang Tong volunteered to return to the Wu troops, assuring Cao Cao that he could arrange for more defections. Sure enough, in due course Cao Cao received a letter from one of the Wu generals, Huang Gai, saying that he was going to change sides and bring with him boats loaded with grain.

The day after the full moon in the eleventh month of 208, Cao Cao's fleet set out to attack. Chained together, it moved as a solid mass. 'When [the boats] got among the waves, they were found to be as steady and immovable as the dry land itself. The northern soldiers showed their delight at the absence of motion by capering and flourishing their weapons.' But what if they were attacked with fire, and needed to scatter, one of Cao Cao's advisers asked anxiously? The leader laughed. The wind is in the wrong direction, he said – if the enemy tried to use fire, it would be blown back onto them.

Seeing the vast armada approach, Zhou Yu was overtaken by a sickness and confined to his bed – an ill omen for the approaching battle. But Liu Bei's military adviser Zhuge Liang came to his bedside and offered a solution. 'To defeat Cao Cao', he said, 'you have to use fire.' But how could that work, the general wondered, knowing what Cao Cao too knew of the wind? Then Zhuge Liang revealed that he had magical knowledge: 'I can call the winds and summon the rains.' He explained that, with a Daoist spell, he could conjure the south-east breeze that was needed to make fire work against Cao Cao.

Meanwhile, the Wu general Huang Gai completed the plan by readying his fireships:

> The fore parts of the ships were thickly studded with large nails, and they were loaded with dry reeds, wood soaked in fish oil, and covered with sulfur, saltpetre, and other inflammables. The ships were covered with black oiled cloth. In the prow of each was a black dragon flag with indentations. A fighting ship was attached to the stern of each to propel it forward. All were ready and awaited orders to move.

Confident of Huang Gai's defection, Cao Cao was unconcerned as the twenty Wu ships approached, despite the south-easterly wind that

Zhuge Liang's ritual had awakened. The latter was nothing to worry about, he told his anxious ministers – of course the wind direction might change from time to time. 'That is my friend, the deserter!' laughed Cao Cao as the vessels drew close. 'Heaven is on my side today.'

But then the trap was sprung:

> When the ships were about a mile distant, Huang Gai* waved his sword and the leading ships broke forth into fire, which, under the force of the strong wind, soon gained strength and the ships became as fiery arrows. Soon the whole twenty dashed into the naval camp. All Cao Cao's ships were gathered there, and as they were firmly chained together not one could escape from the others and flee. There was a roar of bombs and fireships came on from all sides at once. The face of the water was speedily covered with fire which flew before the wind from one ship to another. It seemed as if the universe was filled with flame.

The inferno consumed Cao Cao's fleet. The flames leapt so high, it was said, that they scorched the cliffs red.

The famous victory of Shu and Wu over Cao Cao is far from the end of the tale. The two allies always knew that one day they were likely to face each other in the battle for supremacy; and so it transpired. Zhuge Liang built a fortress at Fengjie to ward off the Wu army, but to no avail. Wu triumphed, and Liu Bei fled to Baidicheng above the Yangtze gorges, where he died. The Three Kingdoms then dissolved into a patchwork of states and would-be minor dynasties, all overlapping and squabbling, until the Jurchen invaders from the north overran Wei in AD 265 and then Wu in 280, forming the precarious (and soon fragmented) first Jin dynasty.

The art of river war

China didn't truly become one empire again until 581, when Yang Jian, Duke of Sui in the Northern Zhou dynasty, seized power and declared the Sui dynasty (see page 116). The duke, now Sui Emperor Wendi, then needed to conquer a southern dynasty called the Chen.

---

* In the combat that followed, Huang Gai was wounded and fell into the river fully armoured. But the fact that he did not drown was taken as 'proof of his natural affinity for water'.

In the 580s, the immense Sui warships defeated the Chen navy on the Yangtze. These five-storey ships were then the largest in the world, holding 800 men and equipped with great spiked balls swinging from derricks. Against this terrifying armada the Chen could do nothing, and for a brief but energetic period the Sui ruled from Guangdong and Hainan to Hebei.

Tall 'tower ships' became a stock feature of the Sui and Tang navies. They are described in the gloriously named manual *Tai bai yin jing* (*Canon of the White and Gloomy Planet of War*), written in 759 by the Tang Daoist and military strategist Li Quan:

> These ships have three decks equipped with bulwarks for the fighting-lines, and flags and pennants flying from the masts. There are ports and openings for crossbows and lances, while [on the top deck] there are trebuchets for hurling stones . . . [The whole broadside] gives the appearance of a city wall. In the Jin period the Prancing Dragon Admiral Wang Jun, invading Wu, built a ship 200 paces in length, and on it set flying rafters and hanging galleries on which chariots and horses could go.

With multiple decks rising to as much as thirty metres, these ships might be armed with 'fending irons': long arms pivoted on jibs and ending in iron spikes, which could be sent smashing down from an upright position to wreak havoc on enemy craft. Meanwhile, swift-moving attack ships known as *meng chong* were used at least since the Han era; the Tang armoured them with plates or sheets of leather, wood, rhinoceros hide or iron, both to give cover from arrows and stones and to repel boarders.

Innovation in naval military technology was one of the most belligerent facets of the inventive 'genius of China' expounded by Joseph Needham in his encyclopaedic examination of how the country's science and civilization co-evolved. A great fleet of warships enabled the Southern Song to fend off Yangtze pirates in the twelfth century, and around the early 1130s a Song official hit on the notion of building ships powered by hand-driven paddle wheels, so that they could be manoeuvred even on windless days. Because their wheels were hidden beneath protective coverings, the ships, called 'flying tiger warships', seemed to the enemies to move by supernatural power, filling them with fear. These vessels had up to twenty-four paddle wheels, but

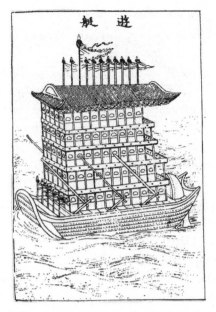

A 'tower ship' depicted in the Qing edition of the *Wu jing zong yao* (*Collection of the Most Important Military Techniques*), originally compiled in the Song era (1044) by Zeng Gongliang. The drawing is based on a description included from an even earlier text, the *Sui shi* (*History of the Sui*, c.636), but Needham says that it is a very unreliable representation, notable mostly for the illustration of the 'fending irons'.

usually just two or four, powered by several dozen crew members. They carried trebuchets that flung gunpowder-filled grenades, and wielded great wrecking balls suspended by chains, or systems of pulleys and booms that allowed rocks to be dropped onto enemy ships from a great height. 'No other civilization produced anything like them', Needham claimed.

Unfortunately for the Song, the bandit leader Yang Yao captured the carpenter who designed the mighty paddle-driven war vessels and forced him to build some for him. By 1135 Yang had a fleet of several hundred with which to defend his piratical activities on the Yangtze. But when, that year, the Song commander Yue Fei fought Yang Yao on Dongting Lake, he devised a strategy to disable the paddle fighters. His troops spread grass and logs on the lake surface, clogging and breaking the wheels. Yang Yao was defeated and beheaded.

That victory did Yue Fei little good in the end. When his heroic achievements began to make him too popular in the eyes of the

Song leaders, the general was imprisoned and poisoned. But thanks to a hagiography written by his grandson, Yue Fei became celebrated during the Ming era as the model of a (wronged but) virtuous servant of the state. There is still a temple dedicated to him today near the West Lake of the former Southern Song capital of Hangzhou, and his slogan 'Recover our Rivers and Mountains' was turned into a patriotic song during the war with the Japanese in the twentieth century.

It's not clear why word of Yue Fei's rather simple strategy didn't get out, but the Southern Song were able to continue using their paddle-wheel ships to good effect in their campaign against the Jurchen invaders – the Great Jin dynasty – in the north. When the two powers clashed in 1161 in the battles of Tangdao (in the East China Sea) and Caishi (on the Yangtze), the technical ingenuity of the Song carried the day. At Caishi the Song commander Yu Yunwen, allegedly leading a force of just 3,000 troops and 120 warships powered by paddle wheels, defeated a Jin navy of 70,000 men and 600 vessels. (The imbalance was almost certainly inflated by the victors' scribes to magnify the achievement.) The Song ships showered the Jin navy with incendiary bombs, a tactic described in 'Hai qiu fu' ('Rhapsodic Ode on the Sea-Eel Paddle-Wheel Warships') by the Southern Song poet Yang Wanli:

> Our ships rushed forth from behind [the island] on both sides. The men inside them paddled fast on the treadmills, and the ships glided forwards as though they were flying, yet no one was visible on board. The enemy thought that they were made of paper. Then all of a sudden a thunderclap bomb was let off. It was made with paper and filled with lime and sulphur. These thunderclap bombs came dropping down from the air, and upon meeting the water exploded with a noise like thunder, the sulphur bursting into flames. The carton [paper] case rebounded and broke, scattering the lime to form a smoky fog, which blinded the eyes of men and horses so that they could see nothing. Our ships then went forward to attack theirs, and their men and horses were all drowned, so that they were utterly defeated.

For all its might and ingenuity, the Song fleet couldn't protect the empire from the Mongol invaders when, after defeating the Jin, they

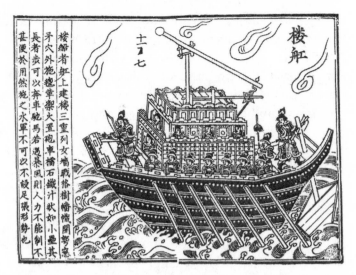

A Song tower ship from the *Wu jing zong yao* (*Collection of the Most Important Military Techniques*, 1044) of Zeng Gongliang. The vessel contains a trebuchet for casting incendiary bombs. This illustration comes from a 1510 edition, and so may correspond more closely to a Ming than to a Song vessel.

turned on their Song allies. Khubilai Khan's cavalry were invincible on the northern plains, but in the south the Mongols needed to fight with ships. They assembled a navy with extraordinary speed, importing sailors and shipwrights from Korea as well as conscripting locals in Shandong. The troops learnt the skills of water combat quickly, and in 1267 they faced the Song fleet at the twin cities of Xiangyang and Fancheng on the Han River – a strategic gateway to the confluence of the Han and the Yangtze – in one of the most celebrated battles in Chinese history.* It was certainly one of the most protracted, allegedly lasting for *six years*, and was fought both on land and on water. The Mongols used their fleet of 5,000 ships to blockade the Han and prevent supplies from reaching the besieged cities, while their cavalry saw off the Song troops attempting to provide reinforcements. Powerful new siege machines such as counterweight trebuchets (a design imported from the Middle East) beat down the city defences. When the Southern Song commander Lü Wenhuan finally surrendered in 1273, the Mongol conquest of China was inevitable. The general Bayan (called Hundred Eyes by Marco Polo, a colourful mistranslation

* Marco Polo claimed to have been there, but that seems unlikely.

of his Mongolian name) battled his way down the Yangtze to the Southern Song capital of Hangzhou, which fell in 1276.

The conquerors were more generous in victory than they were in the winning of it. The Song Emperor Gongdi was a six-year-old boy, and the court was effectively led by his mother, Empress Dowager Quan, and grandmother, Grand Empress Dowager Xie. After the capitulation, mother and son were taken to the northern capital of Khanbaliq, where Gongdi was given the title Duke of Ying. He later moved to the former Mongol capital of Shangdu in Inner Mongolia, and finally to Tibet (then known as Tubo), where he entered a monastery in 1296.

That wasn't quite the end of the Song. A defiant faction of the court escaped with Gongdi's two brothers, and the eldest was declared emperor in Fuzhou, Fujian, in 1276. The entourage was soon forced to flee to Lantau Island, today a part of Hong Kong, where the eldest brother died and the younger, aged seven, was declared Emperor Huaizong. The remains of the Song navy – still a mighty fleet – harboured at Yamen in Guangdong province. In 1279 the Mongol (now Yuan) force, although fewer in number than its opponent, closed in for the endgame, and once again proved its naval supremacy. At the Battle of Yamen the young Huaizong perished along with thousands of officials as they leapt into the sea during that final conflict.

A naval battle ended the Yuan dynasty too: the rebel leader Zhu Yuanzhang, who became the first Ming Emperor (see page 141), crushed the imperial forces, also at Caishi, in 1355. Before he could become emperor, Zhu then had to overcome his rival Chen Youliang, a leader of the Red Turban rebels. The two navies met on Poyang Lake in 1363 in what has sometimes been called 'the largest naval battle in history' (a contested accolade, as you can see). Zhu's ships faced a force three times as great, but he won the battle with his incendiary firepower. Vessels loaded with combustibles, and sometimes with gunpowder, were sent crashing into Chen's triple-decked warships. Chen was eventually killed after breaking out from the lake and being pursued along the Long River.

The Chinese perfected the use of incendiary devices for water warfare, constructing boats that were divided in the middle so that the rowers aft could detach the incendiary fore section and retreat to literally watch the fireworks. Fire was one of the most devastating weapons for river combat, and was developed to a versatile art as naval warfare became ever less a matter of hand-to-hand engagement and more about flinging projectiles. 'Sky-flying tubes' would set fire to enemy sails; 'gunpowder buckets' and 'fire bricks' scarcely need their destructive potential to be spelled out.

If rebels acquired a competent fleet, it was extremely hard for an emperor to suppress them. When the Ming dynasty was in turn superseded by the invaders of the Manchurian Qing empire, one loyalist refused to give up the struggle. The Zheng family of Xiamen (then called Amoy) in Fujian had become so wealthy in its trade with the Dutch that it raised its own private navy. The head, Zheng Chenggong, used these ships to defend the southern coastal regions of Fujian, Guangdong and Zhejiang against the Qing for years. In 1662 the Qing government (led by a regent for the eight-year-old Kangxi Emperor) was forced into the desperate measure of depopulating the entire region of ports and towns within fifty kilometres of the coast to deprive Zheng of resources and support.

Many people of Fujian, seeing their land become empty and desolate, withdrew to Taiwan, which Zheng – now a kind of pirate king

An articulated barge loaded with incendiary bombs, from the late sixteenth century.

who became known as Koxinga – made his own kingdom. (He first had to expel the Dutch, for whom it had become a concession in their alliance with the Qing.) To deprive Zheng of resources and trade, in 1662 the regency ordered that all coastal navigation should cease. 'All ocean-going junks are to be burned', the official edict commanded: 'not an inch of wood is allowed to be in the water.'

Zheng Chenggong himself died in 1662, allegedly in despair at the death of the last Ming pretender in a battle in Burma. But the Zheng family retained control of Taiwan for another two decades, and today Koxinga is worshipped as a god there, and in Fujian too; temples are dedicated to him. His legacy is complex, for almost any party can find reason to claim him as a hero. Since he was born to a Japanese mother in Japan – his father was a sea-faring merchant and pirate – he received the official approval of the Japanese invaders of Taiwan at the end of the nineteenth century. And as his story prefigures the island's status as a refuge of opposition for Chiang Kai-shek's Nationalists in the late 1940s, they too made him a hero. But he was also a useful propaganda icon for the Communists as the man who drove out the Western (Dutch) imperialists and made Taiwan 'Chinese'. The real point is not who has the greater claim to Koxinga's legacy, but that in China the past supplies a multivalent justification for the present.

Gunboat diplomacy

Of course, the truth was that the foreign barbarians were not really driven away by Koxinga at all. Their tall ships kept coming, intent now not on spreading the word of Christ – Christian missionaries were banished from China by the Qing Yongzheng Emperor in 1724 – but on doing trade. They came to an empire possessing riches for which Europeans would pay a fortune but which seemed interested in none of the commodities that the West could offer in return. The result was an alarming trade deficit: porcelain, silk and tea came back to European ports by the shipload, but all that entered China was silver currency.

This would have been intolerable enough. But what truly raised hackles of the Europeans was the supreme confidence of the Qing emperors not only that the West had nothing of interest to offer them but that its peoples were, like China's Asian neighbours, mere vassals.

When George III of Great Britain sent an embassy to the Qianlong Emperor in 1793, led by the diplomat George Macartney, to press for an end to trade restrictions, the emperor made it clear that he had no need of western goods but that nevertheless he expected the British king to strengthen his 'loyalty' and swear 'perpetual obedience'. 'Our Celestial Empire', came the emperor's reply to the king, 'possesses all things in prolific abundance and lacks no product within its borders. There is therefore no need to import the manufactures of outside barbarians in exchange for our own produce.' The message left the British frustrated and offended. They did, however, have one commodity for which there was a thriving market in China. The problem was that it was illegal to sell it there.

Opium, the narcotic extract of the poppy, was long used as a pain-killing medicine, but in the seventeenth century it became popular as what we would now call a recreational drug. Seeing its deleterious effects on the population, the Yongzheng Emperor banned its sale in 1729 and made opium dens illegal. As ever was the case with the drug trade, this merely moved the market underground. Foreign traders in the narcotic continued to sell in China, and in the late eighteenth century the British East India Company, working through private dealers to insulate themselves from the official ban on imports, monop-olized the market of Bengalese opium. So great was the demand that Britain's trade balance with China actually became positive. When the East India Company's monopoly was ended by the British government in 1833, the British expected to be able to continue selling the addictive drug in China as a sovereign power. But the Daoguang Emperor would have none of it, and snubbed the government's chief superintendent of trade, Lord William Napier. Enraged by such treatment, Napier called for British gunships to insist on free trade by force of arms. He died in 1834, but the die was cast. When the Qing official Lin Zexu was dispatched to Guangzhou in 1839 to halt British opium imports, he arrested opium dealers in the province, blockaded the British in their factories, and confiscated and burned their wares without any promise of compensation. Military retaliation was then inevitable.

The First Opium War of 1839–42, while motivated by the British government's determination to push a harmful and socially disruptive drug on a foreign population, was aggravated by cultural misunder-standing. It did not seem to occur to the Daoguang Emperor that either his own citizens or foreign traders and sovereigns might offer any resistance to his moral and political authority. The British officials, meanwhile, were as outraged by perceived lack of respect and neglect of protocol as they were by infringements of supposed trade rights.* When Lin Zexu sent a personal letter to Queen Victoria demanding that she act on her conscience in prohibiting the trade in 'poison', his insolence (worsened by poor translation) seemed in itself almost a declaration of war.

In any event, when the British fleet at Hong Kong (at that time an unpopulated rocky outcrop) drew hostile fire from the Chinese navy in 1839, the Foreign Secretary, Lord Palmerston, ordered that a large naval force be sent from India to teach the Qing Emperor and his officials a lesson.

The war was waged largely on water. While the British troops and navy fought for control of the strategic southern ports and rivers, another fleet headed northwards up the coast towards the imperial city. The Qing government saw the wisdom of reaching a hasty peace agreement, and in early 1841 it agreed to reopen trade and to give Hong Kong to the British. The emperor, however, could not abide the humiliation, and soon the conflict resumed.

On the Pearl River, the key inland route from Guangzhou, the Qing navy proved no match for the heavily armed British fleet, led by the aptly named *Nemesis*, Britain's first iron-clad warship. Once Guangdong was secured, the British forces turned their attention to the Yangtze. They seized Shanghai and moved upriver to Zhenjiang ('Guard the River') at the intersection with the Grand Canal. With this critical junction lost, the Chinese realized the game was up and returned to the negotiating table. In the Treaty of Nanking (Nanjing), signed in August 1842, Britain was awarded concessions and trading rights in five ports, including Guangzhou and Shanghai. More were to follow in subsequent decades. At the same time the Qing officials

---

* Although both parties seemed set on patronizing the other, the condescension of the British can be discerned from the childlike transliterations they imposed on the Chinese language and its pidgin versions of English – 'chop-chop', 'chin-chin' and the like. ('Pidgin' is itself a contraction of the way the Chinese allegedly pronounced 'business' as 'pidginess'.) The former naval captain Thomas Blakiston's remarks about 'wretches' with strange ways and language, as well as his delight and relief in finding pheasants to shoot in China, give a fair indication of the colonialists' cultural sensitivity.

The iron warship *Nemesis* destroys Chinese war junks at Guangzhou (Canton) during the First Opium War. From the *Illustrated London News*, November 1842.

made treaties with the French and Americans, partly in the hope that other foreign interests would keep the British ambitions in check. China began to seem ripe for exploitation by Western colonialists.

The seizure of a British vessel, the *Arrow*, in Guangzhou in 1856 triggered the conflict often called the Second Opium War. Although the ship was flying the British flag, the Chinese authorities accused it of piracy. The *Arrow*'s crew was eventually released, but the British flag was considered to have been 'insulted' by the incident, and Palmerston, now prime minister, saw this as a pretext for strengthening the opium trade and forcing further concessions from the Chinese. Facing both the British and the French plus support from the United States and Russia, the Chinese agreed to the disadvantageous Treaty of Tientsin in June 1858 that opened up more ports to foreign trade and gave the foreigners the right to navigate the Yangtze under a guarantee of safe conduct. But once again the Qing government reneged and elected to fight on.

The ensuing conflict showed why access to rivers and ports was so strategically important. A major focus of the military action was the Bai River (now called the Hai), which flows through Beijing and Tianjin before emptying into the Yellow Sea through the Bohai Gulf. Because Tianjin sits where the Bai crosses the Grand Canal and thus provides access to both the Yangtze and the Yellow River, the river's mouth was protected by five robustly garrisoned forts at a site called Dagu, constructed by the Xianfeng Emperor after the First Opium War. In 1859 the British navy launched an attack on the forts, but even with American help they were repulsed. Not until the French added their might to the assault did the forts fall in August 1860. There was little then to prevent the Western powers from advancing all the way to Beijing, which they entered in October. The old and new Summer Palaces were sacked and looted, and the Forbidden City itself narrowly avoided the same fate. The Xianfeng Emperor fled the capital (and, after his health deteriorated rapidly, died the next summer), but his brother Prince Gong, serving as imperial envoy, ratified the Treaty of Tientsin in late October.

As well as giving foreigners more trading and travel rights within China, the Second Opium War affirmed Shanghai as the new centre of international commerce within the country. It soon became almost a miniature kingdom to itself, with a character unlike anywhere else in China and a reputation for excess and decadence. Isabella Bird, who took advantage of the safe-passage laws to explore China in the 1890s, gave a compelling account of what this former small fishing town had become by then: a city flooded with opium and overawed by Western warships.

> Two big, lofty, white hulks for bonded Indian opium are moored permanently in front of the gardens. Gunboats and larger war-vessels of all nations, all painted white, and the fine steamers of the Messageries Maritimes have their moorings a little higher up. Boats, with crews in familiar uniforms, and covered native boats gaily painted, the latter darting about like dragonflies, were plying ceaselessly, and as it was the turn of the tide, hundreds of junks were passing seawards under the big brown sails.

Thanks to the British determination to peddle it, opium continued to blight Chinese society until the Communists harshly suppressed its use in the 1950s. By the late nineteenth century China was growing its own, making the drug cheaper and more widely available than Indian imports alone would permit. In 1904 one British official reported that half of the urban male population smoked opium, and it was estimated that there were around 20 million addicts. The drug accounted for much of the trade in Sichuan and Yunnan, and the money helped to finance the warlords of the tumultuous 1930s.

The humiliation of China in the Opium Wars indicated to colonialist powers that the country was now up for grabs. The murder of a British diplomat, Augustus Raymond Margary, in Yunnan while returning from an official mission to Burma in 1875 was exploited by the British government as an excuse to demand yet more concessions. In the Treaty of Chefoo in 1876, the Qing confirmed a commitment to safe passage for all foreigners travelling inland, and permitted foreign vessels to sail along the rivers for both military and trade purposes. By the late nineteenth century there were nearly eighty treaty ports, including the Yangtze hubs of Nanjing, Hankou (which merged into Wuhan) and Chongqing. Europeans were free to come and go more or less as they pleased – and the river routes, particularly the Yangtze, were the highways of that traffic.

Releasing the dragon

So much, then, for warfare *on* the waters. But as we saw earlier, military leaders in China sought also to deploy water as an offensive weapon in itself. The flooding of Kaifeng during Li Zicheng's campaign against the Ming was an age-old strategy in China. During the Spring and Autumn period from the eighth to the fifth century BC, the state of Wu in south-east China, occupying the region around modern Jiangsu province and the mouth of the Yangtze, took advantage of its waterlogged geography – and the state's consequent expertise in canal-building and hydraulics – to dam rivers and flood fields and towns during conflicts with the neighbouring states of Song and Zheng. In 358 BC an army from Chu breached the Yellow River as a defensive strategy, using the flood to obstruct the advance of the enemy.

According to the Han historian Jia Rang, dyke-building itself began not for flood control but as a weapon, both to seize water from foes and to discharge its fury upon them. 'The building of dykes', he writes, recently began during the Period of the Warring States when the various states blocked the hundred streams for their own benefit. Ch'i [Qi], Chao [Zhao] and Wei all bordered on the Huang Ho [Yellow River]. The frontiers of Chao and Wei rested on the foot of the mountains while that of Ch'i was on the low plain. Hence Ch'i constructed an embankment twenty-five *li* from the river, so that when rising water approached the Ch'i embankment, it would be forced to flood Chao and Wei. Hence Chao and Wei also constructed an embankment twenty-five *li* from the river [to counteract it].

The ploy of dyke-breaching to flood the enemy was so common, and so destructive, that as early as 651 BC some states signed (to little apparent effect) a treaty banning the breaching of Yellow River levees for warfare.

A particularly detailed account of this practice appears in a document called the *Zhan guo ce* (*Strategies of the Warring States*), a compilation assembled between the end of that period (third century BC) and the Han dynasty of the first century BC. It describes events at the end of the Spring and Autumn period in the fifth century BC, when one of the most powerful of the contending kingdoms was Jin in north-central China, dominated by several elite families. The most ruthless and cunning was the House of Zhi, led by Zhi Yao (or Zhi Bo). He persuaded the Han, Wei and Zhao families to unite with him to destroy the families of Fan and Zhongyang, which they did. Zhi was, however, 'greedy and perverse', and had no intention of stopping there. He now demanded that the earls of Han and Wei cede land to him, to which they reluctantly agreed. But when Zhi Yao insisted on the same from Earl Xiang of Zhao, the demand was refused. Xiang knew what to expect next, and retired to the city of Jinyang (modern-day Taiyuan in Shanxi province) to prepare for assault.

In 455 BC Zhi Yao arrived with his allies Han and Wei and laid siege to the city. He broke open the banks of the Fen River, one of the largest tributaries of the Yellow River, and diverted the waters towards Jinyang. The effects were devastating: the city was flooded up to a height of three storeys, and its defenders had to seek refuge on the tallest rooftops, their kettles hanging on makeshift supports. Amazingly, they nonetheless resisted for three years. But, weakened by disease and starvation – the people were forced, it was said, to resort to cannibalism of children – the Zhao ministers realized that they were all but finished. The earl's chief minister, Zhang Mengtan, proposed that their only hope was to turn Zhi Yao's reluctant allies against him, and he arranged for a secret meeting with the heads of the Han and Wei families. Look, he said to them, do you really believe that if you win this battle, Zhi Yao will honour his promise to give you an equal share of the Zhao territories? Do you not see that you will be next to feel his sword? 'We know it', they admitted. And so they agreed to turn on Zhi Yao.

It's not surprising that they decided to take their chances against the implacable warlord of Jin. While Zhi Yao was out with the two earls inspecting the dykes they had constructed to redirect the Fen during the siege, he had taunted them: 'I did not originally know that water could destroy other people's states', he declared. 'Now I realise that the Fen River can be advantageously employed to inundate An-yi [the major city of Wei] and the Jiang to flood P'ing-yang [in Han].' The earls of Wei and Han understood that some day Zhi Yao was going to destroy them too.

They were not terribly good at concealing their plan to switch sides. Seeing that their allies looked more anxious than elated at the impending fall of Jinyang, Zhi Yao's minister sought to alert his master to the threat of revolt. But Zhi Yao could not believe that this would happen when victory was within reach, and he ignored the warning.

With the complicity of Han and Wei, the Zhao troops crept out from Jinyang and, overwhelming the guards who were minding the breach of the Fen, they destroyed part of the embankments so that the waters flowed towards the Jin army. As the soldiers tried to stem the flood, their allies from Han and Wei set about them. Zhi Yao was captured and brought to Xiang of Zhao, who had him executed and, to expunge the rancour of past humiliations, made his enemy's skull into a lacquered drinking vessel.

There is a colourful postscript to the tale. A former minister of the Fan and Zhongyang families called Yu Rang had defected to the Zhi family before Zhi Yao eliminated them. Hearing of Zhi Yao's fate after Xiang of Zhao had defeated him, the loyal Yu Rang plotted revenge. He hid in the lavatories of Xiang's palace with a dagger, but was discovered and brought before his would-be victim. Yu Rang confessed his desire for revenge, but the earl was impressed by his courage and released him. Reduced to begging, but undeterred, Yu Rang continued to plot the earl's assassination, and he was seized one day lurking under a bridge as Xiang passed by. The earl was understandably perplexed that Yu Rang would be so determined to avenge Zhi Yao when he had not seemed too concerned that the warlord had murdered Yu's former masters Fan and Zhongyang. But they didn't appreciate me, Yu Rang explained, whereas Zhi Yao saw my worth. With a sigh, the earl said that he really couldn't afford to let Yu Rang go free a second time. But the captive declared that he was ready to die; he requested only that he first be allowed to enact a symbolic revenge by cutting up the earl's clothes. With impressive forbearance, the earl allowed him to do so, after which Yu Rang, saying that he was now at peace with heaven, fell on his own sword. That Yu Rang was subsequently mourned and celebrated as a model of integrity, when his only goal was to avenge so unworthy a person as Zhi Yao, perhaps attests to the distance between the morality of ancient China and that of today.

As with most historical tales from those times, the Battle of Jinyang had a moral message. For the third-century-BC philosopher Han Feizi, one of the architects of the harsh political doctrine of Legalism that held sway during the Qin dynasty, Zhi Yao's fate showed what happens when a leader succumbs to greed and perversity. Certainly, the fatal assault by the river waters was engineered by man, not by heaven – but all the same, the implication remained that water will not serve a bad leader.

Zhi Yao's defeat left Jin in the hands of three families, who, by dividing the kingdom into the separate states of Han, Wei and Zhao, initiated the Warring States period. An engineered flood brought that era to a close too. The unification of China under Qin was almost complete when, after conquering the states of Han, Zhao and Yan, King Zheng of Qin turned his attention to Wei. In 225 BC the Qin army, led by Wang Ben, besieged the Wei capital of Daliang. Wang flooded the city by damming a canal on the Yellow River, forcing King Jia of Wei to surrender. Four years later the states of Chu and Qi also succumbed to the might of the Qin, and Zheng became the first Emperor Qin Shi Huangdi.

The tactic of flooding was by this time evidently a routine part of a commander's repertoire. 'Generals confronted by entrenched enemies,' writes military historian Ralph Sawyer, 'whether in fortified cities, freestanding citadels, or amidst fields and mountains, would immediately ponder the possibility of exploiting nearby water resources to inundate them.' This seems only natural in a country with such an extensive water network, but there was more than ruthless opportunism behind the manoeuvre. The use of water as a weapon was suggested by the very character of water itself, as described in the Daoist tradition: yielding yet irresistible, it can wash away the mightiest mountains. The philosophical connection is explicit in the *Wei Liaozi*, a treatise on military strategy attributed to one Wei Liao (tentatively said to have been a Qin minister) at the end of the Warring States period: 'Now water is the softest and weakest of things, but whatever it collides with, such as hills and mounds, will be collapsed by it for no reason other than its nature is concentrated and its attack is totally committed.' Evidently water is also regarded here as a *model* for how military force should be applied: the notion expounded by Sun Tzu in *The Art of War*. 'The strategic configuration of power is visible in the onrush of pent-up water tumbling stones along', Sun Tzu declares. This concept of 'strategic configuration of power', which Sun Tzu called *shi*, is a central component of his military philosophy. He didn't regard the force of water purely metaphorically either: 'Using water to assist an attack is powerful', he advised.

The military manual *Hu qian jing* by the Song-dynasty writer Xu Dong talks explicitly about using the *dao* of water for combat. The Song had bitter experience of that: water was used against them when their enemies the Western Xia in the far north-west steppes burst the embankments of the Yellow River to flood their encampments in 1081 – an attack all the more terrible because the river water was freezing cold. Genghis Khan fared no better in his efforts to defeat the Western Xia in 1209. He tried to divert the Yellow River, in full autumnal flood, to inundate the capital of Yinchuan (now in Ningxia province). But the dykes he built were inadequate, and they crumbled to flood the Mongol camp instead, forcing Genghis Khan to withdraw. (The Mongols returned in the late 1220s, however, when they lived up to their reputation for merciless slaughter.)

Aside from simply drowning the enemy, inducing a flood served other tactical ends. According to the Song compendium of warfare *Wu jing zong yao,* it could turn one's foes 'into fishes, making them live in trees and cook in hanging pots'. Properly directed, the force of the flow could act as a battering ram to destroy fortifications. Or enemy troops could be tricked into crossing a dried-up riverbed, whereupon one would breach the dam that held the river in check, washing the troops away. Using water for war relied on expertise accumulated from quotidian water management. A Tang writer describes how one should use special instruments to measure the water height in order to judge the right moment to 'engulf cities, inundate armies, immerse encampments, and defeat generals'.

The well-timed release of a dammed river flow decided the outcome of the campaign that led to the reunification of China by the Han. The short-lived Qin dynasty collapsed after rebellions led by Liu Bang of Han and Xiang Yu of Chu, who then contended to become the new ruler. In 204 Liu Bang's general Han Xin and his army of 30,000 faced a force of 200,000 from Chu, led by Long Ju, combined with their allies the Qi. They confronted one another along the opposite banks of the Wei River, a tributary of the Hai in Shandong province. The night before the combat began, Han Xin ordered his men to make more than 10,000 sacks and to fill them with sand and stones. With these they fashioned a makeshift dam upstream in the Wei. With the water level lowered, Han Xin marched his troops across to attack the Chu forces – but then beat a hasty retreat. Long Ju had expected no less: Han Xin 'lacks the courage to confront people', he declared. So he instructed his army to pursue the retreating Han forces across the river, at which point Han Xin commanded that the dam be broken down. Many of the Chu soldiers were washed away, and the few who had crossed (Long Ju among them) were now stranded on the enemy bank, where they were easily cut down.

Rather than inundating the enemy, one could simply deprive them of the vital resource. The *Hu qian jing* advised that 'When you want to seize the enemy's strength, first seize their water supply.' This made it imperative for armies to occupy the high ground, the source of the rivers. More terribly, a general might order a water supply to be poisoned. The *Wu jing zong yao* contains advice to defend against that dastardly scheme: 'if the [water's] colour is black and there is foam floating upon the water, or red and tastes salty, or muddy and tastes bitter', then you had better beware. Chinese citizens today would do well to heed this advice in the face of the modern poisoners: the factories and industrial works that squat on the riverbanks, discharging their noxious juices.

Crossing the Yangtze

When flooding induced for military goals got out of hand – as it so often did – the real losers were the ordinary citizens, who died in their thousands or even millions through the folly of generals. Few strategic floods were more catastrophic for the populace than that ordered by the Kuomintang Nationalists in the conflict with Japan in the 1930s.

The Japanese invasion reiterated the strategic importance of the major rivers. From 1937 they drove the Chinese forces steadily back along the Yangtze valley: first they took Shanghai, then (with atrocious barbarism) Nanjing and, by the middle of 1938, the conurbation of Wuhan. Only the precipitous slopes of the Three Gorges stopped their advance, enabling the provisional Chinese government to survive west of the gorges in Chongqing. (It was during his time in Chongqing as director of the Sino-British Science Cooperation Office, an aspect of the Allied war effort in Asia, in 1942–6 that Joseph Needham – until that time a biochemist – developed his fascination with the history of Chinese science, technology and culture.)

By the middle of 1938 the Japanese had taken over virtually all of northern China, but they needed to defeat the Kuomintang army, led by Chiang Kai-shek, at the important railway junction at Zhengzhou on the Yellow River between Luoyang and Kaifeng. From here they could have struck at Xi'an and then south at Wuhan on the Yangtze.

To defend Zhengzhou, the Kuomintang troops were instructed to break down the river dykes downstream at the village of Huayuankou. Ironically, given the difficulty keeping the dykes in place against natural floods, it proved hard to breach them intentionally. The Nationalists tried twice to blast a hole with explosives, before finally resorting to shovels. But once the water began to flow there was no stopping it. The resulting flood covered 23,000 square kilometres, killed around half a million people, and left at least 3 million homeless. As the waters poured away on a new south-easterly course into Anhui and Jiangsu, the displaced population fled into Henan and Sha'anxi, only to face famine there.

Although Chiang Kai-shek tried at first to blame the breach on Japanese aerial bombardment, he was forced eventually to admit that his own troops had caused the devastation. The Communists later exploited that admission relentlessly in anti-Nationalist propaganda. Yet the two sides cooperated after the war to close the Huayuankou breach: a project that, begun in early 1946 with the American engineer Oliver J. Todd as adviser, took a year to complete.

This delicate alliance didn't last long. The two sides signed a peace accord in Chongqing in 1945, agreeing to an orderly reconstruction of the traumatized nation. But hostilities resumed the following year, and the Yangtze was the pivotal locus of the endgame. After almost three years of fierce fighting, in April 1949 the two armies faced one another across the Yangtze. The People's Liberation Army (PLA), under Mao Zedong, was camped on the north bank along the 320-kilometre stretch from Nanjing to Shanghai. The Kuomintang forces were less numerous but better armed, with modern gunboats and aircraft. The river here was wide, swift and treacherous, and seemed to pose a natural barrier against any assault.

Mao conceived a bold strategy. The local people here knew the river well: they were cormorant fishers and they cultivated the marshy banks, growing cinnamon and camphor trees, and mulberry to host silk moths. The Communists won the confidence of the people by expelling oppressive landlords and building schools, so the locals agreed to help them cross the river in a raggle-taggle flotilla of junks, little fishing boats and bamboo rafts. Harking back to Song naval technology, the villagers made hand-driven paddle wheels to propel the boats if

The People's Liberation Army crosses the Yangtze during the civil war in 1949.

there was no wind. PLA soldiers were trained every day to swim and navigate. On 21 April the crossing began under covering fire from the north bank. The first wave carried shock troops, moving as fast as possible to surprise the enemy – although in the event they encountered surprisingly little opposition.

As with many battles in the past, bribery and treachery were used to secure victory. Inducements by the Communists silenced the Kuomintang artillery at the fortress of Jiangyin, where the turncoat Nationalist commander General Tai even directed fire at his own troops. And when the improvised Communist fleet moved on to besiege Zhenjiang, the leader of the Kuomintang warships, Lin Zun, was bribed to hold back his powerful gunboats, so that the PLA took the city with little resistance. This was the turning point of the war. Within two days of the Yangtze crossing the Communists had captured Nanjing, and in October Mao Zedong declared the formation of the People's Republic of China.

8 Mao's Dams

The Technocratic Vision of a New China

Let's wage war against the great earth!
Let the mountains and rivers surrender under our feet.

Zhang Zhimin, *Personalities in
the Commune* (late 1950s)

In 1956 Mao Zedong crossed the Yangtze again. This time he swam it, accompanied by a nervous entourage of officials in Wuhan, and to commemorate the event he wrote a poem called 'Swimming'. The poem announced that Mao's feat had inspired him to an ambitious vision:

I have just drunk the waters of Changsha
And have come to eat the Wuchang fish.
Now I am swimming across the great Yangtze,
Looking afar to the open sky of Chu . . .
Great plans are afoot;
A bridge will fly to span the north and south,
Turning a deep chasm into a thoroughfare;
Walls of stone will stand upstream to the west
To hold back Wushan's clouds and rain
Till a smooth lake rises in the narrow gorges.
The mountain goddess, if she is still there
Will marvel at a world so changed.

China's leaders throughout the modern era have been conscious of how mastery of the country's waters could supply a powerful symbolic demonstration of their right to rule. Heaven might have fallen, but the rivers still held a power of mandate in the popular consciousness. Mao's plan for 'walls of stone' to hold back the waters was not just a component of his programme of modernization through the appliance of science and technology; it was also a demonstration of political will, hatched (the poem implies) during his personal conquest of China's rivers as he swam from shore to shore in the greatest of them all.

Soon after he became national leader in 1949 as chairman of the Chinese Communist Party, Mao identified the Yellow River as a key challenge for the country's promised prosperity. In one of those slavishly repeated but curiously banal slogans at which Mao excelled, he stood on top of a dyke in the autumn of 1952 and announced that 'Work on the Yellow River must be done well.' What he had in mind was the most ambitious project that the People's Republic had yet attempted: not just a programme of flood control that rebuilt and strengthened 1,800 kilometres of dykes, but a staircase of forty-six dams on the river that would hold back the floodwaters and silt.

Through these water-conservancy projects, not only would the problem of flooding at last be solved but the countryside would be abundantly supplied with irrigation. Soil erosion on the loess plateau would be halted with terracing and tree planting, attacking the siltation

Mao Zedong inspects the Yellow River at the site of the future Sanmenxia Dam in 1952.

problem at its source. 'China', the minister of forestry announced in 1956, 'will be a green land.'

It didn't work out that way. Between the 1950s and 1990 more than 80,000 dams and reservoirs were built on China's rivers, but many of these were erected hastily and constructed poorly and have not aged well. A few have fared disastrously.

It would be unfair to characterize the massive hydro-engineering schemes of the Mao era as an abject failure. There were disasters, certainly, but Mao was surely not wrong to imagine that modern engineering might address some of the problems with China's waterways that had proved intractable in the past. The legacy is mixed, and the environmental problems that China faces today are not, on the whole, simply a result of poor water management. It is tempting to ascribe the many shortcomings of the modern water conservancy to bad, overly centralized management, bureaucratic corruption and inefficiency, technical incompetence, hubris and negligence. One can find all these things, but there was nothing new in that – it was the same story from the Sui to the Qing. What is more revealing is how the traditional rhetoric surrounding water management was mobilized for the purposes of constructing a particular vision of modern China. Mao and his mandarins did this because they knew that it was a language the people understood.

Water conservancy in the Republic

Mao Zedong wasn't the first of the nation's leaders who aspired to 'conquer' China's water problems once and for all. Taming the Yangtze by damming was an ambition voiced by Sun Yat-sen after he created the Republic in 1912. But what in the Qing era had been an endless effort depending on massive conscription of labour was now seen as a task to be accomplished by modern technology. Water conservancy had become a science, much of it based on principles developed in the West. Although Sun proclaimed that the establishment of the Republic in 1912 was in the tradition of the French and American republics in its celebration of liberty, equality and fraternity, he was explicit about what the Western model did and did not have to offer China: 'What we need to learn from Europe is science, not political philosophy.'

So the Republic of China had no hesitation in making use of foreign expertise, whether by sending engineers abroad to train or by attracting overseas investment in hydraulic projects. From around 1919, foreign engineers such as the Americans John Freeman and Oliver Todd came to advise on hydraulic technology, while Chinese specialists studied in the West before bringing their skills home. Li Yizhi, often regarded as the 'father of modern Chinese hydraulic engineering', learnt his profession in Germany and toured water projects elsewhere in Europe between 1909 and 1915. For all his technological training, Li retained a somewhat Daoist approach to water management, seeking methods that minimized interference with nature – although that did not prevent him from advocating damming.

Hydraulic engineering, irrigation works and water conservancy are more or less synonymous in Chinese: all may be called *shui li* (水利), literally the profit or benefit of water – you could say the term implies 'making the most of water'. By the early days of the Republic, the objectives of *shui li* had changed. Flood control and irrigation remained priorities, but there was no longer the need for an efficient transport system for tribute grain. Besides, damming the rivers didn't just prevent floods and create reservoirs; it produced a potential new source of power, hydroelectricity, to drive China's economic expansion. This represented a change in attitude: China's water resources were not simply a problem to be managed, but were part of a development plan that amounted to nothing less than the rebirth of Chinese civilization in the modern age. In 1924 Sun Yat-sen announced in a speech in Guangzhou that

> If the water power in the Yangtze and Yellow rivers could be utilized by the newest methods to generate electrical power, about one hundred million horsepower might be obtained . . . When that time comes, we shall have enough power to supply railways, motor cars, fertilizer factories, and all kinds of manufacturing establishments.

The need for this resurrection was urgently felt. After the social and environmental chaos of the late nineteenth century, with the Yellow River abandoned to do its worst and the North China Plain ravaged by flood and famine, China presented a sorry face to the world. The Republic was a fragile state, and between 1916 and 1926 it was all but impotent as the country was carved up by warlords. Diluvian legend provided the obvious metaphor for China's plight. In 'The Age of Flood' ('Hongshui shidai', 1921), the poet Guo Moruo drew a parallel between the tide of Western imperialism and anarchy of the warlords and the struggle of Da Yü to rescue the nation from inundation and impose order. 'How can I face my people', Yü exclaims in Guo's poem, 'if I cannot control the flood?'

Although Sun Yat-sen served only very briefly as (provisional) president of the Republic in 1912, he continued to lead the Kuomintang during the chaos of the warlord period. Sun sustained an alliance with the newly formed Communist Party to suppress the rebels and separatists. But after his death in 1925, leadership of the Kuomintang was assumed by the rightist Chiang Kai-shek, who cracked down harshly on the Communists when they seized control of Shanghai, fuelling the tensions that soon exploded into civil war.

Chiang announced a bold plan to tame the Huai River with dams and conservancy schemes. Here he carefully selected a goal that was impressive without being unachievable. Ever since the Qing government had ceased trying to manage the waters by abolishing the Yellow River Administration, the Huai (which was included in the administration's responsibilities) had suffered. The river has always been a strange one, lacking any clear route to the sea. Depending on changes to the course of the Yellow River, it has sometimes found a wandering path to the coast, sometimes made its way into the Yangtze, and sometimes it pooled into Hongze Lake and other swampy lakes. What's more, being on the boundary of the northern and southern climatic zones it experiences extreme weather variability. So the Huai has always been difficult to manage and prone to disastrous flooding. All the same, it represented an economically important target for water conservancy without presenting a challenge of the same order as the Yellow River or the Yangtze.

In 1929 Chiang established the Huai River Conservancy Commission (HRCC), headed by Li Yizhi. Despite a major flood of the Huai and Yangtze in 1931 and a lack of funds occasioned by the war with the Communists, the Japanese occupation of Manchuria and other drains on finances, the commission more or less completed its plan to enhance the Huai's drainage into the Yangtze and dig a new outlet route to the sea. (Chiang undid all this hard work with the disastrous flooding of the

Yellow River into the Huai basin in 1938, described in the previous chapter.) All the same, the HRCC created the tradition of a centralized, scientifically and technologically oriented bureaucracy, informed by international expertise, that the Communists inherited and – except during the isolation of the Cultural Revolution – have mostly sustained ever since.

In the march of hydraulic nation-building, the Yellow River could not be ignored. Quite the contrary: as the Republican government began actively to weave the Yellow River into a narrative about the origins of Chinese civilization, it became ever more symbolic of a mandate to rule. The Yellow River Conservation Conservancy Commission was created in 1933, headed by Li Yizhi and advised by specialists such as Todd, who advocated damming schemes for flood control and hydroelectricity.

The problem of the Yellow River became something of an international cause célèbre. A solution could win its architect universal acclaim. Top German hydraulic engineers such as Hubert Engels devoted their energy to the problem: Engels constructed a large model of the river at his research institute at Obernach in Bavaria. Yet it wasn't clear that the conventional wisdom gained from experience with European rivers could be transferred to the unique difficulties posed by the Yellow River's immense silt load. 'The Hwang Ho [Huang he] and other rivers of North China', a League of Nations report concluded in 1935, 'represent hydraulic phenomena, which, in intensity and size, go beyond – frequently far beyond – anything of the same kind observed on Western rivers.' This suited the later rhetoric of Chinese exceptionalism: when damming of the Yellow River began in the 1950s, the newspaper *Renmin ribao* claimed that only the Chinese truly understood the river and that all previous efforts had been tainted with unwelcome political agendas.

Mao's dam-building

In 1992 I visited the ancient Buddhist site of Lingyin Temple near Hangzhou, at the entrance to which stands the craggy peak of Feilai Feng. The limestone is pitted with grottoes and caves, their walls lined with Buddhist statues housed in little alcoves. Many, I noticed, lacked a head, and I asked the guide why. In retrospect it was a little unfair to put the unfortunate young woman on the spot like this: she was forced to mumble the obvious pretence that they had been damaged in a flood. The truth, of course, was that they had been beheaded by Revolutionary Red Guards, intent on honouring Mao's 1966 call to reject the 'Four Olds': old customs, culture, habits and ideas that had poisoned people's minds for thousands of years. But maybe the guide was right after all, for the flood unleashed by the Cultural Revolution was arguably more devastating for China than anything the great rivers had produced.

Quite aside from the wanton vandalism it triggered, the Maoist critique of old traditions exposes the hypocrisy of a regime otherwise content to milk the past for propaganda. (But which regime, communist or capitalist, has not done that?) When Zhou Enlai announced plans for harnessing water on the North China Plain in 1949, he was well aware of the resonance in proclaiming that 'Our achievements will be nothing less than those of Da Yü.'* The Communists reinstated the Yellow River Conservancy Commission along the lines of the Nationalist model, and the first large-scale project of note was the People's Victory Canal, a network of irrigation channels stretching for 200 kilometres between the Yellow River and the Wei. It was completed in 1952, when Mao was on his iconic tour of the North China Plain. The Communist Canal and Red Flag Canal were later added to this system. The latter, a 1,500-kilometre channel completed in 1966, relied largely on local resources and expertise rather than centralized bureaucracy, and became seen as a paradigm for the power of patriotic mass labour to enable a socialist-style modernization.

In the 1950s some experts argued that such modestly scaled local initiatives, aimed primarily at supplying irrigation water but which could also reduce flood risks, were the most effective model for water management. Some of these projects were realized, and some were useful. But the Great Leap Forward saw potentially productive labour commandeered for empty shows of industry. Just as pots and pans were melted in countless backyard furnaces for pig iron that no one needed, so irrigation channels were dug for no purpose save the symbolic exhibition of communal enthusiasm for state policies. And in the same way, this pointless activity was actually counterproductive, disrupting

---

* It may be thanks to Zhou Enlai that there is anything left at all at Lingyin, incidentally, for he urged restraint on the Red Guards who might otherwise have razed the whole place.

well-established rhythms of life and exhausting the workforce. The grain output on the North China Plain was lower in 1959 and 1960, after the irrigation campaign, than it had been on average before. What *did* increase was the water withdrawal: between 1957 and 1959 the volume of irrigation water tapped from the Yellow River rose by an astonishing eighty-three times, without any agricultural benefits. Indeed, the inexpertly made irrigation systems did more harm than good, leaving soils waterlogged and eventually causing problems of salinization.

During the Great Leap Forward, every county was told to build dams and diversion channels. Most projects lacked the requisite design and construction skills, and even if materials of adequate quality were available (often they weren't), inferior ones might be used anyway to impress the party chiefs with cost savings. As a result, many of the structures lasted barely a couple of years. Even during the Cultural Revolution, errors of judgement were officially admitted: Zhou Enlai said in 1966 that

> I fear that we have made a mistake in harnessing and accumulating water and cutting down so much forest cover to make way for more agricultural cultivation. Some mistakes can be remedied in a day or a year, but mistakes in the fields of water conservancy and forestry cannot be reversed for years.

He was right, but it was already too late.

The wall falls

China's dam construction has been viewed with such mixed feelings, if not outright scepticism, over the past half-century that it is not said often enough why the principles behind it are largely sound. For a country with such extensive water resources and so great a demand for energy, it would seem foolish not to harness the flow of the rivers for hydroelectricity. And dams are surely a valid means of retaining water for irrigation (they had been used for this purpose on a very modest scale since the Three Kingdoms era of the third century AD), and for controlling the catastrophic floods that have blighted the river valleys. Whatever one might think of China's mega-engineering schemes, doing nothing is not an alternative.

The key problem for the Chinese state has always been the same one. It's a problem of scale. The country is so vast, its climate so extreme, its needs so boundless, and its water courses so immense, that nothing but engineering of epic proportions seems likely to make an impact. And if the preference for large-scale hydraulic engineering was surely ideological to some degree, it was also in tune with the orthodox view of river management worldwide in the immediate post-war decades. But the bigger the project, the bigger the challenges, and the greater the risk and cost of failure.

Like the Nationalists before, the Communist government began large-scale damming not on the Yellow or Yangtze rivers but with the lesser challenge of the Huai. No one could claim that the problems of the Huai River were solved under the Republic, and after major floods in 1949 and 1950 Mao Zedong grappled with them again, launching a new campaign to 'Harness the Huai' for water control and hydroelectric power. Now there was no question of appealing to the Americans for advice – if anything, resisting the rivers offered a metaphor for the struggle against Western imperialism. It was the Russians, now, to whom the Chinese looked for help. At first this meant struggling with Russian engineering textbooks; but after the Sino-Soviet Treaty of Friendship, Alliance and Mutual Assistance of 1950, Russian engineers began to arrive to offer their experience in person.

Like most damming projects, the campaign on the Huai had its critics. Some Chinese hydrologists warned that increasing water retention in the river basin would make agricultural land waterlogged and alkaline. But such warnings went unheeded, and dam-building proceeded on a large scale throughout Anhui and Henan. Administrators sometimes overruled their 'conservative' engineers, demanding that the dams be designed to hold as much water as possible so that they could offer the greatest boost to local irrigation and agricultural productivity. One of the engineers who objected was Chen Xing, an acknowledged expert who had designed the massive Suya Lake Reservoir in Zhumadian prefecture, Henan. He was denounced as a 'right-wing opportunist', and dam construction forged ahead.

Dams need to be designed to withstand the worst that nature will throw against them. But in East Asia the worst is hard to predict. Tropical cyclones brewed up on the fringe of the Pacific Ocean may trigger once-in-a-lifetime weather conditions that engineers might be tempted to overlook. That's what happened in the summer of 1975: in early August, barometers plunged in the Philippine Sea as the cyclone that would be christened Typhoon Nina began its fateful gyration. It moved steadily west and swept across Taiwan, causing floods and landslides and destroying thousands of homes. Taiwan's mountains weakened the hurricane, but it was still a furious force when it made landfall in mainland China at Fujian province. It progressed north-west across Jiangxi and Hunan, but was halted by cold air at Zhumadian. With nowhere else to go, the storm hung over the valleys between the Funui and Tongbai Mountains and discharged its rain.

The downpour was reportedly like a firehose, powerful enough to kill birds in flight. A metre of rain fell in three days, which was extreme even for this storm-prone region. The water streamed down the slopes into the Ru River, a large tributary of the Huai that runs through Zhumadian City. There it built up in the reservoir behind the Banqiao Dam, built in 1951–2 to help tame the Huai.

The Banqiao had a 116-metre wall made of clay. It had been constructed with the aid of Soviet engineers, who had advised on strengthening the wall when cracks appeared shortly after the building was finished. These repairs, engineers insisted, had turned the Banqiao into one of the allegedly indestructible structures known as 'iron dams'.

The Banqiao engineers can't be accused of neglecting the possibility of freak weather: they had designed it to cope with the kind of extreme rainfall predicted only once in 1,000 years. Unfortunately the statistics on which they based this estimate were poor, indicating that such a once-in-a-millennium event would precipitate just over half a metre of rain in three days. The remnants of Typhoon Nina delivered twice that.

As the rain thundered down throughout Henan, smaller dams gave way to the inrush of water. By 8 August, dozens of them had collapsed, and the mighty Banqiao had become a source of huge anxiety. The sluice gates were opened to try to discharge the rising mass of water, but they couldn't do it fast enough, partly because they were obstructed by sediment. As the water level reached the top of the embankment, workers were dispatched along it to repair the damage that had started to appear.

Like many natural disasters, the collapse of the Banqiao Dam around 1 a.m. on 8 August acquired its own mythology. There was a deafening peal of thunder, it is said, after which a 'hoarse old voice' was heard to cry, 'The river dragon has come!' It came in the form of 700 million cubic metres of water, released from the shattered clay barrier in a torrent six metres high and eventually reaching twelve kilometres wide, coursing through the dark early morning at about fifty kilometres per hour. Entire villages and towns simply vanished under the wave. Some of those close to the dam had been evacuated already, but residents who lived further away were not so lucky. The fatality figures are inevitably disputed. One early estimate, which is probably conservative, put it at 85,000, although this was later officially revised to 26,000 by including only those who had died directly from drowning, not deaths due to the disease and famine that followed. Some estimates today are as high as 230,000.

In any event, the consequences of the Banqiao collapse are hard to disentangle from those of related failures. On the Hong River, which converged with the Ru at Huaibin, the Shimantan Reservoir Dam collapsed just half an hour before the Banqiao, initially releasing over 25,000 cubic metres of water per second. Where these two flood surges met, Huabin City was more or less obliterated. As the waters rushed down the river valleys, other dams downstream were intentionally destroyed by air bombing to try to discharge the great wave in selected directions. But because flood diversion and dredging schemes in the region had been neglected for decades, the floodwaters eventually just sat there, leaving survivors stranded in trees and on rooftops for days or even weeks, desperately hoping for food or rescue as the rains were replaced by intense summer heat. Dysentery, typhoid, hepatitis, malaria and other diseases claimed many more victims than the floodwaters had. By 19 August, nearly half a million people in Zhumadian remained stranded.

Convention demanded that someone be blamed. In November, Qian Zhengying, head of the Ministry of Water Resources and Electric Power, accepted 'responsibility for the collapse of the Banqiao and Shimantan dams', saying, 'We did not do a good job.' Qian wasn't entirely a scapegoat; her ministry shared some culpability, for their safety procedures and plans for dealing with a crisis of this magnitude were clearly inadequate. But the real failures were systemic and political. There were no mechanisms for ensuring that dam-building was done well, nor to guard against incentives (embezzlement of funds, eagerness to please authorities, exaggerated claims of efficiency and effectiveness) for doing it poorly. By 1973, 40% of the largest dammed reservoirs – those with capacities of up to a million cubic metres – failed to meet the standards specified for them at the outset and were beset with sedimentation problems. There were on average over a hundred dam collapses a year since the campaigns of the Great Leap Forward: nearly 3,000 by 1980, and over 500 in 1973 alone. It's hard to avoid the suspicion that the leaders had assimilated the message on which China's rivers had long insisted: life in their shadow is cheap.

Taming the Yellow Dragon

In the late 1950s Mao's government launched an almost frantic campaign of damming on the Yellow River. Forty-six structures were built along its length to restrain the flow of silt. The newspaper *Renmin ribao* spoke of battling the river in the way that legends told of bouts with the river dragons: the engineering projects would 'chop off the scales, claws and teeth of the wicked dragon'.

The greatest of the Yellow River dams was at Sanmenxia ('Three Gates Gorge'), where the splitting of the river into three main channels by a group of small outcrops offered the opportunity to span the mighty river in manageable stages. Construction began amidst great optimism in April 1957, with the help of Soviet engineers. The new reservoir meant that thousands of peasants in the valley had to be relocated: an exigency that soon became familiar in river valleys all over China. The claim was that the dam would retain silt to avoid the old problem of a rising riverbed on the plains downstream: the Yellow River, it was argued, would be yellow no more. It was an absurd, utopian belief, but it fitted an ancient prophecy: 'When a sage appears, the river will run clear.'*

The Soviet advisers withdrew after 1960 as relations between the two Communist powers soured, and they became convenient targets for shifting the blame when siltation proved to be a bigger problem than anticipated. Or, perhaps, just bigger than had been admitted. At the start of the Sanmenxia construction in the spring of 1957 it was permissible for engineers such as the American-trained Huang Wanli to voice worries about the silt, which the Yellow River water carried

---

* Popular usage of that idiom showed more wisdom: 'when the river runs clear' was the Chinese equivalent of 'when hell freezes over'.

Murals of the Yellow River at Sanmenxia before and after the construction of the dam.

in sobering quantities of around thirty-six kilograms per cubic metre at the dam site. At a meeting that June, Huang asserted that fantasies about making the waters run clear 'distorted the laws of nature'. He recognized that a grandstanding project like this had little hope of succeeding without first carrying out less glamorous measures to control erosion and silt ingress further upstream.

But it soon became dangerous to voice such objections. That month, Mao decided that 'letting a thousand flowers bloom' might threaten the ability of the party to push forward with its plans, and so all opposition to official decisions risked being labelled 'rightist'. Huang suffered for the criticisms that he had earlier been encouraged to express: accused of having sympathies with foreign countries and bourgeois democratic ideals, and of having attacked Mao personally, he was stripped of his professorial post at Qinghua University and sentenced to hard labour on a construction site, before being sent to the countryside for re-education. He suffered beatings and other ill-treatment, and his children were victimized and forced to denounce him.*

Nature, unconcerned whether Huang's criticisms were rightist or not, proved him correct. Silt began to accumulate against the dam wall at an alarming rate, and by 1962 the capacity of the Sanmenxia reservoir was almost halved. It was decided that the silt should be removed after all, and the reservoir had to be emptied

* Huang was reinstated as a professor at Qinghua University in 1980, but remained an outspoken critic of China's mega-engineering projects, including the Three Gorges Dam and the South–North Water Diversion Project.

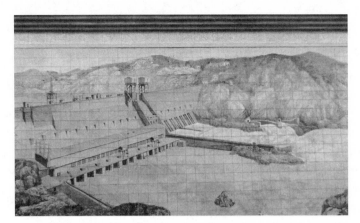

and dredged, while tunnels were constructed (at great cost) to sluice the sediment around the sides of the dam during the flood season. Needless to say, this meant that the original aim of silt retention had to be largely abandoned. Even after extensive reconstruction between 1965 and 1973, almost 40% of the dam's capacity to retain silt was used up in the first eighteen years of operation, and so much water had to be allowed to pass directly through the wall that the hydroelectric power generated by the dam was far less than (barely 5% of) the original estimates. Efforts to improve the structure were hampered by the anti-intellectual climate of the Cultural Revolution, during which educated technical experts were ridiculed and removed from office. In 2004 one of the engineers involved in the design of Sanmenxia, Berkeley-trained Zhang Guangdou, admitted on Chinese television that the dam had been 'a mistake'.

That is before one even considers the problems of resettlement. Sanmenxia was one of the first cases of forced mass relocation in the Mao era, as 280,000 peasants were moved from their lands in the Yellow River valley upstream of the dam in Sha'anxi, Shanxi and Henan to the remote provinces of Ningxia and Gansu. Despite official reassurances, they found that little thought had been given to how they would eke a living in those bleak and barren lands. Some, finding that the problems with the dam meant their former homes were not flooded after all, had to petition for years to get them back, and many did not do so until the early 1980s.

Today the Sanmenxia Dam continues to do its job in desultory and inefficient fashion, with the air of a bored old soldier remaining forgotten at his post after the conflict has moved on elsewhere. A few bewildered tourists, sheltering under umbrellas from the harsh Henan sun, wander with disconcerting freedom among the rusting hulks and stained concrete of the Great Leap Forward era while the ageing power wires hum persistently in the background. There is really nothing at stake, and in truth this lends a kind of affecting charm to this neglected outpost of Maoist wishful thinking that contrasts sharply with the brash posturing of the Three Gorges Dam on the Yangtze. A statue of Da Yü stands guard over the parched hillsides, but one feels sure he would have done things differently; as Huang Wanli's daughter asserted, 'Of all China's leaders, only Yü the Great managed the Yellow River well.'

Some engineers have long acknowledged that the river's hazards have been worsened considerably by many centuries of deforestation on the loess plateau, leaving soils exposed to rapid erosion by rain. But efforts to solve this problem have generally been poorly executed. One can't rush reforestation – but in the face of impossible demands for a quick fix, that's exactly what work teams ended up doing.

The Sanmenxia Dam on the Yellow River. The slogan on the dam wall says 'When the Yellow River is at Peace, the Nation is at Peace' – a saying attributed to Da Yü.

Environmental engineering simply can't be done well when the priority is to please superiors, to reach ambitious targets faster and more cheaply, and to create a veneer of excellence adequate only to ensure political preferment. In the first several decades of Communist rule the stakes were often too high, the rewards too great, the penalties too severe. And so trees planted in haste failed to take root, and did little to anchor the yellow earth of Sha'anxi and Shanxi.

Even by the 1980s there was no real improvement to the rates of erosion in the Yellow River catchment. Some farmland simply vanished. In 1972, for example, the entire village of Chenjiagedu in Sha'anxi upped and left, since there was no longer any soil fit for cultivating. That's the harsh lesson of this inclement environment: clear the land for farming, and soon enough there will be no land left to farm.

Soil loss and sediment accumulation remain serious problems for the Yellow River basin today, nearly two-thirds of which continues to suffer excessive erosion despite schemes to replant forests and grasslands. The Asian Development Bank estimates that the attendant problems cause annual economic losses approaching 3.5% of China's GDP. The losses for the farmers eking out a precarious livelihood are, of course, far greater.

Efforts to control the Yellow River have not by any means been futile, however. From the early 1950s the dykes were reinforced with stone and strengthened by planting of willows, and the flow rate and erosion at bends in the river were reduced by building outcropping 'spur dykes'. When there were heavy rains in 1958, no serious flooding ensued. Today, those living in the river valley are safer from floods than they have ever been.

Yet the real achievements of Yellow River management are not easy to discern behind what has often been inflated propaganda. The broader picture is to be discerned not in facts and figures but in the matter of general strategy. Preventing erosion, or installing modestly scaled flood controls in the remote headwaters of the great rivers, is the kind of mundane, low-tech but capital-intensive river management that has tended to hold little appeal for government agencies and ministers. That doesn't mean it hasn't happened; small-scale, local irrigation and flood-management projects have had some success. But the ambitions of the state have veered towards the kind of grandiose, gargantuan and prestigious projects of which Sanmenxia was an early example – even when some engineers and other experts have warned of their limited value. These initiatives arguably make the 'hydraulic state' model more relevant to the People's Republic of China than ever it was to imperial times. The symbolic significance of these massive, costly enterprises is evident from the way that they have been actively constituted in the light of China's water mythology.

Making a new myth

As Aldous Huxley and George Orwell understood, the success of the totalitarian regimes of the twentieth century depended not just on repression and control of the populace but on the active construction of national myth. Here the state leaders adopt the role of wise, courageous patriarchs who bravely resist decadent and predatory foreign influence, while the self-sacrifice expected of citizens is celebrated with heroic images and narratives. The state becomes the source of moral virtue.

China's massive hydraulic engineering projects were an excellent fit to this rhetorical mould. In older times they were shaped by Confucian ideals of selflessness and duty, which were rather easy to accommodate into Communist ideology. Participation, once compelled by force on a voiceless peasantry, now became a collective expression of the will to build a cohesive society. Imperialism had failed to manage the waters well because it lacked moral authority and worked for the benefit of a few; by succeeding, socialism would confirm its mandate. As an official document on the Yellow River strategy in 1959 put it, 'Accomplishing this pioneering engineering feat will manifest the unarguable superiority of the basis of Marxist–Leninist theory as [the] guiding thought of the socialist system.' (Again, Western politicians have no cause to sneer at such sloganeering – replacing 'Marxist–Leninist' with 'free market' and 'socialist' with 'capitalist' makes for no less fatuous a claim.) By dominating nature, the party becomes an inexorable and incontestable force: it becomes the mountain, imposing its will on the flow of water.

Whatever the official line about the 'Four Olds', Mao understood the valorizing power of historical precedent in China. In 1958–9 he called for a reassessment of the reviled first Qin Emperor Shi Huangdi; in the 1970s he and the prominent officials known as the Gang of

Four encouraged a more favourable view of the Qin Legalists who had argued for powerful, authoritarian rule from the top down.* As the historian John Keay puts it, 'history and culture served as the currency of [political] debate and suffered greatly in the process'.

Other inevitable casualties in this process were the maturity and integrity of media reportage. The stories that embellished the postures of heroic water management in the Mao era have a fairy-tale naivety. When the People's Victory Canal was opened in 1952, *Renmin ribao* reported how peasants rushed to ladle out the water and taste it, and finding it 'sweet' and rich in fertile mud, leapt through the streets with joy.

A state-endorsed book, triumphantly titled *Conquering the Yellow River* (1978), by Wei Huang, gives a fair impression of the message that was fed to the Chinese people at this time. It is a sobering, even chilling reminder of how history was routinely and grotesquely rewritten into a fabulist celebration of Chairman Mao Thought, peppered with bold-face quotes from the Chairman himself:

> Long, long ago, the people of China's various nationalities began to live and work in the [Yellow River] basin, transforming nature and creating wealth. Through class struggle and production, they contributed towards creating China's magnificent ancient culture . . . Like the river itself, the popular revolutionary struggle surged tempestuously ahead, sweeping away one reactionary dynasty after another, and pushing China's history forwards.

This stuff is easy to mock, but it must have been deeply painful to read for anyone who cared about the country's history.

To justify the martial attitude of defeating nature, the Yellow River itself became an emblem of political oppression, 'inflicting frequent disasters on the people'. As Deng Zihui, vice premier of the State Council, explained in 1955, 'In the past, the scourge of the Yellow River was inseparable from the crimes of the reactionary ruling class.' That was intended not as metaphor but as a cause-and-effect relationship: as Wei writes, 'Throughout history, the reactionary ruling

* With his strong assertion of state power, the current president, Xi Jinping, seems to be continuing this rehabilitation of the Legalists.

classes paid no attention to harnessing the river or using it properly'; they (and they only) merely used it as an instrument of war (witness the Ming governor defending Kaifeng against rebels in 1644) to 'shore up their cruel rule'.

But there *was* symbolism in this image too: in the broken, flimsy and hole-ridden dykes, one could read the state of the ruling class who neglected them. The worst negligence was attributed to the Communists' most recent enemies, whose memory and legitimacy it was all the more essential to efface. And by implying that the past disasters of the Yellow River basin were linked to the iniquities of a feudal and reactionary past, engineering projects that tame the waters became a symbolic way to banish those injustices. Only the Communists, then, have tried to use the river responsibly and 'for the well-being of the people', and only they have any chance of success. Here, competence in water conservancy is once again being used as a gauge of fitness to rule.

To motivate the Herculean effort required, Mao alluded to the legend of the Foolish Old Man who moved the mountains (see page 105). But new folk tales were also needed to support and stimulate the effort, and Wei Huang eagerly provided some. During the construction of a dam in the Qingtong Gorge in Ningxia, he wrote,

> A peasant in his seventies came on donkeyback after a ten-day journey from his home village in the Liupan Mountains area in south Ningxia, an area with a glorious revolutionary tradition, carrying cold rations. He pleaded to join the construction, and said stroking his beard, 'It's only under the leadership of Chairman Mao and the Communist Party that we people of the various nationalities are able to channel our energies for the good of the coming generations.'

Conquering the Yellow River endorses Mao's picture of a technological defeat of nature (see Chapter 10) – a transformation of the landscape into a modern industrial complex that guarantees prosperity. Wei's descriptions of this marvellous future send a chill down the spine as they foreshadow all too accurately what is to come:

> New houses, schools, stores, smokestacks and other factory buildings can be seen all along the Yellow River . . . now the people of Inner Mongolia have changed the name of the place from 'land of deer' to

'land of iron and steel' . . . [In Henan,] In addition to building new homes for themselves, the farm workers have set up a number of small enterprises including a machine repair works, a phosphate fertilizer plant, a power plant, a paper mill, a brick kiln, and various types of processing plants.

At least this aspect of the modern mythology for the new state was, alas, grounded in reality.

To hold back Wushan's clouds

Sun Yat-sen's dream of a dammed Yangtze tantalized every subsequent state leader as the ultimate symbol of authority. Sun had imagined a series of dams in the upper reaches that would put an end once and for all to the Long River's floods and provide abundant hydropower. In 1933 the Nationalists had planned a thirteen-metre dam in Xiling Gorge, near Yichang, but the cost was prohibitive. And in 1944, with the Japanese still occupying much of eastern China, the American engineer John Savage, who had worked on the Hoover Dam, came to inspect the Three Gorges region for the Chinese government. (The Japanese had even drawn up their own damming plans for this region in 1940, which informed the ideas of the Americans and Chinese after the plans were seized.) In October, Savage presented a proposal for an immense 225-metre dam. President Roosevelt endorsed the project shortly before his death, and the United States agreed to give China a loan of $3 billion to build it. But after the end of the Second World War the plan was shelved, and then abandoned, because of resumed hostilities between the Nationalists and the Communists.

Mao's poetic evocation of a Yangtze dam in 1956 was motivated in part by the wish to prevent the kind of massive flood that, two years earlier, had killed 30,000 people. At that time he had appealed to the Soviet leader, Nikita Khrushchev, for assistance, and by the time Mao wrote 'Swimming', the Chinese and Soviet engineers were already arguing over where to build it.

At first the choice seemed obvious. In the Three Gorges, the 200-kilometre stretch from Yichang in Hubei downriver to Baidicheng, the river was confined to a narrow channel that reduced the distance a wall must span – in places it was just 105 metres from bank to bank. That squeezing sped the flow so that the river water was able to scour one of the deepest beds in the world. This swift, deep current carried a vast amount of energy that could be tapped for hydroelectric power. What's more, the steep ravines made any large-scale human settlement impossible within the gorges themselves. Nowhere further east of the Three Gorges (it was largely in the east that hydroelectric power was needed) presented such an inviting situation.

Yet even if the Three Gorges region was the best candidate, it wasn't clear exactly where in this long corridor a dam should go. When in the 1930s the Nationalists engaged American engineers to identify the best site, they advocated a bend near Sandouping, a small fishing town in the prefecture of Yichang. The plans advanced no further during the war, and when the Communist Party came to power American involvement was no longer welcome. All the same, Mao's committee agreed that Sandouping looked the most promising location – until his generals cautioned that this relatively wide stretch would be vulnerable to air attack. With armed conflict so recent in memory, that seemed an important consideration.

As the deliberations dragged on, they became increasingly politicized. Officials became more concerned to second-guess the 'correct' party line and to avoid 'reactionary' positions than to assess the evidence objectively. As a result there was a herd-like hopping from one site to another, only for each to be rejected as inappropriate due to the cost, geology or whatever. After Zhou Enlai led the arguing parties on a tour of the Three Gorges in 1958, it was eventually agreed that a 200-metre dam would be built at Sandouping after all. But these plans languished in the tribulations of the Great Leap Forward and the Cultural Revolution, and it wasn't until 1970, when the political turmoil began to subside, that the great dam on the Yangtze returned to the agenda.

The first firm decision was to construct a pilot project at Gezhouba, forty kilometres downstream of Sandouping. More than a 'test run', the Gezhouba Dam was intended also to ease the challenges of the Three Gorges Dam itself. Construction began, with much fanfare, on Mao's birthday in 1970. It was initially envisaged as another 'people's project', carried out in true revolutionary spirit by an untrained horde of volunteers without any real leadership, plan or expertise. Within two years of commencing, however, the construction had become so chaotic that it had to be suspended until a proper planning authority was instituted. Lin Yishan, who headed the Yangtze Valley Planning

Office (YVPO) since its inception in 1956, was placed in charge. At any rate, Gezhouba turned out to be a poor monument to the Great Helmsman, failing to deliver anything like the amount of hydroelectric power first promised and taking eighteen years to finish at a cost some three times the original estimate. Only as the construction was nearing completion could the Three Gorges Dam return to centre stage.

China's new leader, Deng Xiaoping, was a hydraulic engineer, and he looked favourably on heroic schemes. Besides, the middle and lower Yangtze valley was still desperately short of power in the 1980s, putting a brake on economic growth; hydroelectricity looked like the ideal solution. In 1982 Deng announced his support for a dam at Sandouping – and with the reforms and opening up of the Chinese economy, the way was now clear again for foreign investment. The financiers and engineering conglomerates of the international dam-building community – Merrill Lynch, the US Army Corps of Engineers, the World Bank – recognized a lucrative venture. With interest from Canada, Sweden and Japan too, the Three Gorges Dam looked set to become an international project.

But Li Rui, Mao's former vice minister of electric power in the 1950s and now a vice minister at the Ministry of Water Conservation, opposed the plans. He felt that flood control would be more effectively achieved with a series of smaller dams, and that such a large under-taking was beyond China's economic means anyway. Li had spent twenty years in prison as an 'anti-party' element during the Cultural Revolution; released and rehabilitated in 1979, he was a formidable and determined critic.*

The dam's leading advocate, Lin Yishan, was no less committed to his position. He was no slavish ideologue; although he lacked formal training in water management himself, Lin had assembled a capable and informed staff for the YVPO. Li and Lin had clashed over the dam project back in 1958, when Mao had demanded that they each present their cases to him personally. On that occasion Li came off better, and Mao's lack of enthusiasm stalled the project over the next

---

* Li Rui has continued to express profound disagreements with the party line. He was at one time Mao's personal secretary, but has denounced the former leader's dictatorial governance and the callousness of his policies. He has called for political reform, greater press freedom, and a readiness to face up to the past.

two decades. Lin had been sidelined until he was placed in charge of the Gezhouba construction.

But in the 1980s it was now Lin who gained the upper hand. All the same, Li Rui was not alone in his objections. The politically powerful Li Peng, another hydraulic engineer, was also an early critic, although his reservations seemed to wane after Deng made him vice minister of water conservancy and power in 1982. In 1984 Li Peng was elevated to vice premier of the State Council; four years later he became its premier. By the time Li eulogized the dam project in 1992, it had plainly become as much about China's international prestige and power as about riparian management. 'The Three Gorges dam', he said, 'will show the rest of the world that the Chinese people have high aspirations and the capacity to successfully build the world's largest water conservancy and hydroelectric power project.'

The plan that Li Peng announced was awe-inspiring: a wall 189 metres high (the water level would reach to 175 metres) and 2,092 metres wide, requiring 26 million tonnes of concrete and 250,000 tonnes of steel. The reservoir lake, covering about 1,000 square kilometres of land, would stretch back for almost 600 kilometres, past Chongqing, which would then be in a position to receive 10,000-tonne ocean-going ships.

The stakes were now very high. If the Three Gorges Dam were *not* to be built, China's apparent lack of national resolve and resources would have been exposed on the global stage. Failure was therefore unthinkable: as Premier Zhu Rongji warned the construction teams in 1999, 'any carelessness or negligence will bring disaster to our future generations and cause irretrievable losses'. It had become impossible to disentangle the practical aims of the project from its symbolic status. It would be a source of national pride, a glorification of the state. Never mind rhetoric about the supremacy of Communist thought; the Three Gorges Dam would represent China's emergence as a world superpower.

The waters rise

Debate about the Three Gorges Dam has generally been more polarized outside of China than within. With the machinery of the state driving it forward, it has sometimes seemed that any suggestion of technical flaws was brushed aside with high-handed dismissiveness.

What about the forced relocation of thousands of people, or the effects on local ecology, or the loss of sites of priceless national heritage? Blithe reassurances from the Chinese government that these would all be dealt with have failed to create much international confidence.

It is hard to navigate the merits of these arguments, not least because critics of the dam have had a tendency to adhere to every imaginable objection while advocates insist that there are no problems at all. Yet there is surely much truth in the judgement delivered by the political scientist Lawrence Sullivan:

> All the ingredients familiar to bureaucratic machinations are here: promises of administrative position and influence, buying-off of potential grass-roots opponents, unrealistic budgetary figures that seriously underestimate true costs, 'smoke and mirrors' methods of financing, glossing over and even ignoring potential dangers while exaggerating the benefits, falsifying data and fixing crucial technical experiments, and bureaucratic manipulations to ensure that viable alternatives never come up for discussion, let alone a decision.

While the Three Gorges Dam was simply a project under discussion, its strengths and weaknesses could be debated in China without endangering one's career (or worse). But once the scheme had official approval, objection became more perilous. It did not help that the key decision-making stage came just after the Tiananmen Square incident in 1989, when the Chinese government was anxious to suppress all hint of political dissent. At that time, one of the most vocal opponents of the dam, the writer Dai Qing, was jailed without trial in a notorious maximum-security prison in Qincheng. She was held in solitary confinement for more than half of her ten-month sentence, and was told at one stage that she would be executed.

Dai was adopted as a child by the one-time minister of defence, Marshal Ye Jianying, after her father, Ye's friend, was executed by the Japanese in 1944. She trained as an engineer in Harbin, and was a staunch patriot working on missile weapons systems until the Cultural Revolution showed her a different face of Mao's agenda. Her 'third father', a scholarly man named Tang Hai who had acted as Mao's translator in the 1930s, was imprisoned after falling out with the chairman and died in an asylum. Dai's mother, who had suffered appallingly under the Japanese while pregnant with Dai's sister, was also imprisoned and tortured. As intellectuals, Dai and her husband were sent to the countryside for 'reform through labour', and their daughter was taken from them and given to a peasant family to raise; she did not see her mother again for three and a half years.

Despite her disillusion and disgrace, Dai was eventually rehabilitated and in the early 1980s she became a government spy, posing as a writer. That occupation became much more than a cover, however: she proved to be an able novelist and journalist, and began writing for the intellectual newspaper *Guangming ribao* (*Enlightenment Daily*). In 1986 Dai covered a small conference organized by Chinese scientists concerned about the Three Gorges Dam, and she collected her interviews into the 1989 book *Yangtze! Yangtze!*; when it was published by an obscure publisher in Guizhou it was promptly banned and Dai was imprisoned. Her follow-up collection of essays, *The River Dragon Has Come!* (1997), met the same fate, and sealed Dai's reputation as a troublemaker.

Dai's story illustrates the contradictions, baffling to outsiders, of life in modern China. One might imagine that her imprisonment would be considered an indelible stain, but on the contrary after her release in 1990 she was able to resume her career as a writer and was even permitted to travel abroad, visiting Harvard and Columbia universities. She resists calls for hasty democratic reform in China (she had attempted to persuade some of the Tiananmen student protestors to leave before the violence erupted), and seeks instead an intermediate path of gradual and cautious change. Yet she is prohibited from publishing her works or obtaining official employment in China. When Dai was invited to appear at the 2009 Frankfurt Book Fair, Chinese officials threatened to boycott the event and the invitation was withdrawn.*

Curiously, the history of Dai's arch opponent, Li Peng, in some ways mirrors her own. Li too was orphaned in the 1930s – his father was executed by the Kuomintang – and was adopted by a prominent

---

* The affair became a scandal. Dai attended anyway as an unofficial visitor of the German branch of the international writers' rights group PEN. The fair's organizer expressed displeasure at the censorship, and one of the managers was sacked for ruling that Dai and the dissident poet Bei Ling should not speak in the closing ceremony.

member of Mao's government, none other than Premier Zhou Enlai. Li also trained as an engineer, majoring in hydroelectric engineering in Moscow. When he became premier, Li seemed to be the archetypal Chinese technocrat, convinced that massive engineering projects held the key to the country's economic future. The lack of technical training in political leaders is often rightly lamented in the West, but such knowledge can also be a dangerous thing: for politicians like Li, China's water problems will be solved only by gargantuan technological intervention, not by looking at questions of management or allocation.

Li initially hoped that the Three Gorges Dam would be completed in 1997, to coincide with the return of Hong Kong from British to Chinese rule. But things did not go so smoothly. In 1992 he expected that approval for the commencement of construction would be granted as a mere formality at the National People's Congress. Yet when the chairman of the meeting prohibited any criticisms from being voiced, one-third of the delegates protested by voting against the plan – not enough to block it, but a level of dissent unheard of at the congress, which was widely seen as a vote of no confidence in Li himself. Foreign investors got cold feet – the tide of opinion was in any case turning against giant dams, which were seen as too costly, environmentally problematic, and prone to siltation. By 1994 China was having to organize and finance the project by itself.

Nothing, however, could stop the juggernaut now. The first phase of construction began that year, when an artificial channel was dug to divert the Yangtze so that the riverbed could be dried in preparation for the pouring of the foundations. The project proceeded apace, as Sandouping was transformed into an industrial complex managed with military efficiency (and, indeed with military personnel). Towns and cities in the valley upstream were first evacuated, then demolished. For a time, the low-lying districts of bleak Yangtze cities such as Fengjie and Wuzhou took on a post-apocalyptic appearance, where remaining residents eked out a living amongst piles of rubble.

By 2003 the vast wall of concrete was in place and the reservoir began to fill, raising the water 135 metres. The state news agency, Xinhua, proclaimed the triumph with a Maoist flourish:

Is this not a vivid expression of the powerful cohesion of the Chinese nation and the superiority of socialism in pooling strength to accomplish great things?

It's tempting to dismiss this kind of thing as crude, bombastic propaganda, not least because it is trivially false: big dams had been built elsewhere without socialism to guide them, and such massive engineering projects can equally be achieved with the support of, say, capitalistic private investment or military-inflected interests of democratic governments. But to deride the rhetoric is to miss the point. There is every reason to believe that these are sincerely held sentiments, since they draw on centuries of experience of large-scale hydraulic engineering made possible only by the ability of a strong centralized power to mobilize vast manpower resources. If China makes the mistake of thinking that statements like this one are perceived as anything but hokum in the world outside, the *waiguoren* are equally mistaken to think that they are nothing but empty postures of ideological supremacy.

The drowned and the displaced

As the waters began to rise, opposition to the dam had already moved to a different sphere. By the early 2000s, veteran campaigners like Dai Qing were largely ignored within China. The majority view, according to the Asian-affairs specialist Deirdre Chetham, was that the Three Gorges Dam was necessary and inevitable. For that very reason, it was no longer particularly dangerous to voice dissent: the carefully stated concerns of some intellectuals and engineers would be tolerated but ignored.

For the past decade or so, these concerns have focused not on trying to halt operation of the dam but on ameliorating the problems it might create. Some say that the retention of sediment will worsen the flood risks upstream in Sichuan. They worry that the change in river flow and the creation of a massive reservoir will impair the ecology and environment of the region. They worry about the robustness of the immense shiplocks on the river, and indeed of the dam wall – could a major earthquake nearby bring it toppling? There are fears that the sheer mass of the reservoir might itself heighten the earthquake hazard, if not for the dam then for surrounding regions. The weight of the water, as well as infiltration into the pores and cracks of rock beds, might trigger tremors in regions already seismically active. There have been suggestions that the magnitude-8 quake that hit Wenchuan county in Sichuan and killed around 70,000 people in 2008 might have been caused by the filling of the Zipingpu Reservoir near Dujiangyan on the Min River, completed in 2006. Connections have also been asserted between the magnitude-6.5 earthquake in Ludian county, Yunnan, in 2014, with a death toll of several hundred, and two dam reservoirs on the Jinsha River: the 286-metre Xiluodu Dam at Yongshan, forty kilometres from the Ludian epicentre, and the 161-metre Xiangjiaba Dam, serving the two largest Chinese hydroelectric power stations after the Three Gorges. Tremors at Ludian seem to have become more frequent at the time the dams were completed, around 2012, although such a correlation remains far from providing convincing proof of the link. The Three Gorges Dam does appear to have increased the incidence of minor tremors, triggering landslides in the region that temporarily displaced villagers into tents and tunnels. Areas at risk of subsidence now pepper Fengjie county.

One of the most controversial issues is the permanent resettlement of those whose homes were flooded within the reservoir basin. The area designated for the Three Gorges Reservoir – covering 1,000 square kilometres, stretching more than 600 kilometres upstream and, unlike the Great Wall (contrary to the popular myth), visible from space – was home to around 1.5 million people, living in nineteen counties and municipalities, 140 towns, 326 townships and 1,351 villages. All needed to be relocated.*

Would those affected be given enough money to start a new life elsewhere? How would peasant farmers find the skills to survive if they were rehoused in urban environments? Would the moves exacerbate crowding in a region already said to exceed its population-carrying capacity? And what would be the psychological impact of being forced off land on which your family might have dwelt for generations, and where respect for the buried remains of your ancestors is of utmost importance, not least for ensuring the welfare of future generations?[†]

---

* In 2007 it was announced that a further 4 million people would need to be resettled in the region by 2020. Although around half of these are in the reservoir area, officials in Chongqing municipality insisted that the resettlement was part of development plans for Chongqing itself and not dam-related: it represents an attempt to shrink the economic gap between urban and rural populations.

[†] The sense of personal connection to an ancestral home is reflected in the way that the Chinese character *jia* (家) can equally mean 'home' or 'family'.

These concerns need to be put in context. For one thing, while it is tempting to imagine an idyllic way of life on the river that was disrupted by this harsh intrusion of state megalomania, for many of those resettled there was little to celebrate in their situation before the waters rose. Deirdre Chetham describes typical Yangtze river towns in the 1980s as 'chaotic collections of stucco and sod houses in the hills, grim, gray apartment blocks next to factories spewing unrecognizable, boiling liquids into the river.' These places, says Chetham, already suffered a 'widespread exhaustion of spirit . . . a kind of weariness that tends to prevail in places where opportunities are few, nothing is clean, and there is a sense of decay and decline'. We might then be less inclined to share the surprise expressed by writer Peter Hessler at the attitudes he encountered during the two years he spent teaching in the Yangtze town of Fuling in Sichuan in 1996–7, which would lose some of its lower reaches to the Three Gorges Reservoir. There was, he says, little apparent concern among many of the residents, including some who would have to move, about the impending inundation of Fuling's lower slopes. Those who would not be directly affected seemed to feel no solidarity with the (often poorer) Fuling residents who would be. Peasants who were losing their fields seemed quite happy to be given one of the new apartments on higher ground at discount prices, along with the financial compensation which seemed to them to be a great deal of money. Perhaps they were being blinded by cash settlements that, while offering more money than they'd ever possessed before in one lump, would not last them long without further employment. But as Hessler pointed out, peasants from the countryside were already migrating in huge numbers to the towns and cities in search of work; to do so with a cheap apartment and a wallet full of cash would understandably seem a good deal. In any event, rather few Chinese share Westerners' instinctive affection – perhaps one might say sentimentality – for old towns, particularly when the alternative is a nearby modern apartment with amenities that they have never enjoyed before.*

---

* I have heard of a Chinese visitor to London who complained that the English people seem very lazy because they cannot be bothered to bulldoze the old parts of the city and replace them with something more up to date.

Besides, as Li Boning – deputy director of the Three Gorges Dam Construction Committee and the leader of the resettlement plans – has argued, at least half of those to be resettled already lived in towns rather than the countryside, and so they could resume the same employment as before. Many of the townships, like Fuling, would only be partially flooded anyway, and so the displaced families could be shifted simply to a new part of town rather than having to move long distances.

Yet if there was rather casual acceptance of a situation that seems unthinkable to Westerners used to the securities of private ownership, the reasons may go deeper. Resettlement was nothing new – it had been happening to make way for hydraulic engineering projects for hundreds of years, at least since the construction of the Grand Canal. During the Mao era, people had been moved around China at the whim of the state, perhaps simply because Mao decided it was good for them: such disruptions served his purpose for breaking down traditional social structures and ways of life in order to cultivate a revolutionary mentality. As Hessler put it:

> The truth is that the disruption of the dam, which seems massive to an outsider, is really nothing out of the ordinary when one considers recent history in the local context. Within the last fifty years, China has experienced Liberation, the radical (and disastrous) collectivization of the Great Leap Forward, the Cultural Revolution, and Reform and Opening . . . In comparison it seems a small matter to turn a river into a lake.

There is of course a danger here of dismissing concerns and dangers just because they seem less distressing than the traumas that went before. But the fact remains that, for people accustomed to being assigned jobs in a different city from their spouse and a long way from children left in the care of grandparents, resettlement caused by civil engineering may not seem so out of the ordinary. It does appear that the objections to resettlement raised by the dam's critics are not always shared by the people they are speaking for. When Hessler put the question directly to some of his educated colleagues, they found it an odd one. In the two years he lived in Fuling, Hessler wrote, 'I never heard a single resident complain about the Three Gorges Project, and I heard gripes about virtually every other sensitive subject.'

This seems to be a reflection neither of naivety nor of fatalism, but, rather, of the distinctly Chinese accommodation between private and public concerns. A survey of 1,000 people in the Wanxian region who would be affected by the dam, conducted by researchers at the University of Alabama at Birmingham in 2007, found that most participants were optimistic about general abstract questions regarding the dam – 97% agreed that it would bring benefits to the region and the country. But they were less hopeful about their own circumstances, anticipating loss of income and worse living conditions.

It was much the same story about the loss of cultural heritage. The Three Gorges are full of history, but some of it is now lost, probably in perpetuity, beneath the reservoir water. In the typical Chinese style of moving the mountain, some ancient sites and buildings were simply relocated brick by brick. In 2003 the Zhang Fei Temple in Yunyang county, built in memory of Liu Bei's general in the Three Kingdoms period, was moved to a site with similar *feng shui* to the original, ten kilometres away and further up the slopes. Similarly, the ancient Qin-era town of Dachang in Wushan county was allegedly 'cloned' (into something more like a theme park) in 2006, while the original is now 'Dachang Lake'. Baidicheng, made famous by Li Bai, Du Fu and Liu Bei, is said to be unaffected: the official Three Gorges guidebook insists that it has become 'a beautiful peninsula' where 'a beautiful lounge bridge [whatever that might be] [will] connect it to the north bank of the Yangtze'. The remarkable Shibaozhai fortress built into the cliff face near Chongqing will be protected from danger of collapse following the submersion of its mudstone base by 'slope protection and steel reinforcement'. Yet however well these measures perform, there's no denying that many sites of archaeological significance are gone for good.

If China often displays a lack of concern about its cultural heritage, this is perhaps not so different from the way Europeans in the past would dismantle old churches and monuments to build their houses, or rebuild ones deemed too old-fashioned. Even in the mid-twentieth century, China was so steeped in old relics, buildings and artefacts that these items could scarcely acquire much prestige solely on the basis of their antiquity. Only now, when so much of the past has been destroyed or actively rejected, does one increasingly hear regrets about the cultural loss. Even then, the brutal truth is that a part of the motive for preserving ancient sites comes from their commercial value, especially for tourism. But ensuring that this heritage remains intact comes low on the list of priorities of many people. One of the chief engineers on the Three Gorges project dismissed worries about the loss of ancient structures and artefacts below the reservoir lake on the basis that 'the common people of China have such a low educational level that they will not be able to enjoy these cultural relics'. There is surely some cynicism in this remark, but it also reflects a lack of idealization about the past or the 'simple life' that comes from a culture all too aware of the grim realities of those situations. For many Chinese people, modernization, despite its faults, is hugely preferable to what went before – and it is easy to see why they would think so.

How have the resettled migrants fared? The only fair assessment is that their experience has been mixed. For that very reason, it is possible for objectors to find some desperate cases where resettled people have been placed in dire circumstances with insufficient compensation, sometimes exacerbated by thoughtless, inefficient or actively corrupt bureaucracy (one estimate suggested that more than 7% of the resettlement budget had been embezzled by 2002). Government officials, on the other hand, do not have to look too hard to find model cases in which things have gone according to plan and the migrants are happier and more comfortable for it. What's more, these are early days: it is inconceivable that such a massive displacement of communities would have happened without hitches, and some of the challenges seem likely to be temporary. For example, many migrants spoke only their local dialect fluently, which is incomprehensible to their host communities hundreds of miles away; but the second generation of migrants will very quickly adapt to the local lingua franca.

The outcome for migrants has often depended on minute particulars: what the local economy and management is like, what conditions were agreed for resettlement, and what personal and financial resources they have to draw upon. No one would claim, for example, that working the land for self-sufficiency in a poor region around the Three Gorges was easy labour, and yet for a fifty-year-old it might have been less physically and mentally exhausting than working long hours in a packing factory. Some migrants have been able to start up new businesses even if they lacked previous experience – to become butchers or run a restaurant, say. Others don't have the business acumen, or the luck, needed to succeed.

So the issue comes down to so many thousands of case histories, from which it is hard to get a clear picture. Take the case of Shanghai, which was compelled to accept 7,500 migrants from the Three Gorges region between 2000 and 2004 (officials from the Shanghai region were among those who voted against the dam project at the unruly People's Congress in 1992). For migrants sold on the promise of a mythical booming Shanghai known only through television, the reality was sobering. They were all assigned to rural districts remote from the city centre, such as Jiading to the north or the island of Chongming in the Yangtze delta, which has expanded through sedimentation into the third largest island in China. On the other hand, the conditions of re-settlement seemed to contain some generous provisions: everyone would receive a single-storey house and one *mu* (about 700 square metres) of cultivable land, along with (capped) free medical care and two years of free schooling for children. Even the seemingly minor provision of a concrete pathway to the house was an attraction for villagers used to slogging along muddy tracks to their door in the rainy season.

But the houses were cramped and often shoddily made. It was all very well being able to grow your own food and allowed to sell any surplus, but the cost of living even in the outskirts of Shanghai was frightening compared to that in the Hunan countryside. Once education began to cost after the two years of grace, it was hard to find the money, despite the great value Chinese people place on schooling. On Chongming, where the economy is primarily agricultural, jobs are scarce and the new migrants faced stiff competition from others who had come from the countryside as far afield as Sichuan, lured by the glamour of the big city. Perhaps worst of all, the official scheme of resettling families in groups of just two or three households distributed among the Chongming villages, where transport was sparse, broke up the social networks that could have offered migrants support. So despite having access to water, toilets and bathrooms, electricity and health care that they could not have found in the hills and mountains around their native towns of Wanzhou and Yunyang, some of the migrants have struggled to integrate and hesitate to say that their lives have improved. Those on Chongming were among the most challenged, but many elsewhere had personal stories of hardship too. One Mr Chen in Jiading was settled securely enough, but he had been forced to leave his mentally impaired brother behind in the care of neighbours. He had hoped that bringing his brother to Shanghai would be a formality once he was there himself; he had not counted on the intransigence of the authorities.

It seems almost disrespectful to wrap up such difficulties into any blanket statement about the pros and cons of displacement. But perhaps in that respect Mr Chen's predicament represents the right way to think about all the pains and problems of China's modernization: namely, in terms of the lives of individual people, coping as best they can with a process that no one fully understands and to the dilemmas of which there are no easy solutions.

The Three Gorges today

So the dam is built, the reservoir has filled, the temples are flooded and the people resettled. It is the most tremendous and awe-inspiring dam project on earth: 185 metres high, almost two kilometres wide, and flooding 30,000 hectares of agricultural land to create a reservoir the size of Lake Superior. The dam already has an iconic status comparable to the Great Wall, and features on a bas-relief in the Millennium Hall in Beijing.

It is also a tourist attraction, and the (mostly Chinese) visitors seem drawn by a mixture of pride and simple curiosity. The tours impose a state-approved experience: you are ferried by electric buggy between vantage points calculated to showcase the overawing scale of the

The Three Gorges Dam at Sandouping today.

engineering. Huge chunks of the construction machinery are put on display, weathering as though they are already historical artefacts. In such ways a visit to the Three Gorges dam speaks eloquently – I am sure inadvertently – about the continued dialogue between water, power and authority. The guidebook proudly lists the world records set by the dam, many of them meaningless to all but the most avid disciple of hydraulic engineering: 'the highest intensity of concrete placement', 'the spillway dam with the largest discharge capacity', 'the inland shiplock with the highest total water head'. A visitor can hardly fail to feel dwarfed by the scale of such a project – and one can't help thinking that this is the explicit intention. This is too big an affair for you, the onlooker is told: you can only marvel, pay your respects – and put your trust in the state machinery capable of such a feat.

Regardless of one's personal views about the Three Gorges project, it would surely be folly for a country like China, with plenty of fast-flowing rivers and an immense demand for energy, not to make use of its natural renewable resources to sustain economic growth. Around 1,000 megawatts are being added to China's coal-plant generating capacity *every week*, and the environmental consequences of mining and burning this dirty fossil fuel are onerous both for the local populations and for the world. China theoretically has more potential hydroelectric resources – around 380 gigawatts, equivalent to hundreds of medium-sized nuclear power stations – than anywhere in the world, but is still exploiting barely a quarter of that. The government's laudable aim that by 2020 15% of the country's power will come from renewables and that its carbon emissions will be reduced by 40–45% only seem feasible if hydroelectricity is a big part of the mix. (The Three Gorges Dam alone is said to be capable of satisfying more than 10% of the national energy demand, although it remains to be seen if that figure is realistic.)

On top of this, dams do have the potential to alleviate problems of flooding, to provide irrigation water and improve river navigability. There can be no denying the gains: the Yellow River, for example, has not suffered a major flood since 1949. Today the complex system of dams and sluice gates on the river is controlled by a computerized system linked to monitoring stations and satellite observations: the so-called 'Digital Yellow River'. This scheme can be used to flush away sediment and pollution spills, has led to some improvements in water quality, and has reduced the over-exploitation that saw the river's lower reaches repeatedly drying up in the late 1990s (see Chapter 10). In 2010 the Yellow River Conservancy Commission was awarded the prestigious Singapore-based Lee Kuan Yew Water Prize for exemplary water-resource management.

Yet big dams cause problems – for ecosystems and the environment (the carbon footprint of making a large dam should not be underestimated), for water availability and international tensions, for resettled populations and for public health and safety. There is now widespread scepticism among experts that immense dams are the most effective way of harnessing a river's bounty. In retrospect, the planning of the Three Gorges Dam looks like the last gasp of an old-style planned economy, in which the money and workforce are assumed to be limitless and obedience unquestioning. In the end, the decision was forced through not so much because no one would admit that the plan might have problems but because that is the way the Chinese government was used to working. Yet even if it can claim some degree of success, the project also revealed how outmoded and counterproductive this mode of governance is. In the modern age of global science and information, it is becoming ever harder to exaggerate your achievements and hide your failures.

9 The Fluid Art of Expression

How Water Infuses Chinese Painting and Literature

Under my feet the moon
Glides along the water
Near midnight, a gusty lantern
Shines in the heart of night
Along the sandbars flocks
Of white egrets roost,
Each one clenched like a fist,
In the wake of my barge
The fish leap, cut the water
And dive and splash.

Du Fu, 'Brimming Water'

Reading the Tang poets Du Fu and Li Bai, widely considered the finest China has ever produced, you sense that water was the prevailing element of daily life in those days. Take Du Fu's series of poems on the great floods and rains of AD 754: they leave you feeling damp and muddy, oppressed by dark skies heavy with raindrops:

Wind of rain, lurking rain
Autumn flurrying turmoil,
Seas and wastelands circling the earth
Share a single cover of cloud
Horses going, oxen coming –

One no longer can tell them apart.
The muddy Ching and clear Wei
When again can we distinguish them?

The capacity of water to represent the vicissitudes of human nature and fortune finds eloquent expression in verses like these. Du Fu and Li Bai (also called Li Po) are often dubbed the River Poets, and one might fairly assert that water is their primary metaphorical vehicle. They both spent years wandering in the valleys of China's two great rivers, and those mighty waterways rarely seem far from their thoughts. Du Fu's style was itself considered to embody the characteristics of water: the Song scholar and poet Huang Tingjian called it 'like the highest mountain or the deepest water, seemingly unreachable'.

Water works its effects in these poems with exquisite subtlety, connecting the images just as a stream provides visual continuity while tumbling through a landscape painting – visible here, hidden there. 'A dog barks amid the sound of waters', writes Li Bai at the beginning of a poem with the koan-like title 'Visiting the Recluse on Mount Tai Tian and Not Finding Him In' – and the sound of water implies the traveller's immersion in nature from which the dog's bark awakens him. This same water is glimpsed throughout the rest of the poem: here it has carved a ravine, there it raises the 'blue haze' of a waterfall, here it bears peach blossoms back to the world, to which the traveller too must ultimately return.

To the inexperienced reader of such works (I do not pretend to be anything else), the water-drenched imagery might be easier to discern than the aesthetic quality. With good reason has it been said that reading classical Chinese poetry is as much of an art as writing it. Poetry in translation always misses a great deal, of course: the rhythms and rhymes, allusions and stylistic flourishes all tend to suffer. But that is truer than ever for Chinese poems, where the visual appearance of the characters and calligraphy must be sacrificed entirely, along with the density of meaning, quotation and implication that the highly homophonic character of the Chinese language makes possible. Couple this to the extremely compressed literary form – four characters may do the work of as many sentences, and literal translation is mute to the latent meanings – and you can safely say that many of these poems demand so much exposition for the uninitiated that their artistic qualities are at risk of being thoroughly obliterated.

If that sounds discouraging, rest assured that, whatever infelicities translation might impose, and whatever the blindness of the novice to the contextualizing references of the Tang literati, the works of Li Bai and Du Fu retain their heart-stopping poise and elegance, their sense of irony and melancholy. They are still able to transport us to the riverbank where, under dripping boughs, we sit with the tipsy author and read life's travails in the relentless passage of the water. Even if rain today no longer threatens to bring a bad harvest and famine, a flooded countryside, or even a cold and uncomfortable walk home, we can readily enough appreciate those realities for the poet. Even if China's rivers today are lined with factories spewing effluent, and the waters are strewn with noxious garbage, we can sense their cool, sweet embrace for the weary traveller in the heat and dust of a Tang summer. We can understand why, as they looked into the surging Yangtze or Yellow River, or gazed over the serene expanse of Dongting Lake or Hangzhou's West Lake, these artists could apprehend the narratives of the nation's history.

And if it takes many hours, weeks, years fully to decode these poems, that seems a fair price to pay for the time and effort alleged to be necessary for composing them. Before a word was written or a brushstroke was made, the poet and the artist had to contemplate and imbibe nature. According to the Ming writer Dong Qichang, a painter should 'read ten thousand volumes and walk ten thousand miles':

> All these will wash away the turgid matters of the mundane world and help form the hills and valleys within his bosom. Once he has made these preparations within himself, whatever he sketches and paints will be able to convey the spirit of the mountain and rivers.

Some of the imagery is familiar, even commonplace. Many cultures have made a stream or river a symbol of time – of change and transience, but also of longevity and permanence. Poets and philosophers were humbled by the knowledge that the great rivers had flowed from their source to the seas since long before they walked on earth, and would do so long after they had gone. Empires come and go, Chinese writers were fond of declaring, but the rivers and mountains remain. What passes, Confucius attested, is like a river: 'day and night it never lets up'. There is consolation in the thought. As a thirteenth-century Song general mused as he sat in retirement on Beigu Mountain in Zhenjiang (a peak celebrated in the *Romance of the Three Kingdoms*, and therefore redolent of times past):

> How many dynasties have risen and fallen
> In the course of long centuries
> And history goes on
> Endless as the swift-flowing Yangtze.

What poets found in water was not so much the wonder of nature's sublime beauty but a reflection of their inner world, their hopes and regrets. Time and again the river heralds departure and separation. Here is Li Bai, the notorious drunkard, leaving a dear friend at a wine shop in Nanjing:

> Go and ask the Yangtze, which of these two sooner ends:
> Its waters flowing east or the love of parting friends.

And here again wishing farewell to another companion:

> Your lonely sail a distant gleam that fades amidst blue hills;
> The River's course to far horizons all that I can see.

Quoting these stanzas on his Yangtze voyage, the Song official Lu You commented that 'That gleam of sail and mast against distant hills is particularly admirable, and not something you can understand if you haven't travelled for a long time on the River.'

Water could do much more than symbolize yearning, loss and sorrow – and not only for Li Bai, not only for poets and writers, but for artists of all description. We've seen how in China it has aspects both philosophical and political; it speaks to the highest aspirations and most abstract ideals, and it is rooted in daily life and in the exercise of power. Those polyvalent qualities have made it indispensable to the artist, not just for personal expression but as a vessel for subtle commentary and criticism. This is as true today as it was twelve centuries ago.

When water features in Chinese art, we need to attend closely. It is probably telling us more than it seems.

Wandering and exile

The two most famous Tang poets are typecast as having contrasting personalities and capabilities. Li Bai was the romantic, drunken vagabond with inimitable, innate talent; Du Fu was the earnest Confucian scholar with impeccable technique. These are at best sketchy caricatures; in any event, the two men were not rivals but friends. Both seem to have been restless and thwarted, and often close to poverty. Li Bai's family origins are obscure (those he asserted may well be invention), and it seems possible that they lay outside China in Central Asia, from where his family moved to Sichuan when he was young. The truth is all the more obscured by Li Bai's tendency towards self-mythologizing: he asserted, for example, that he had been a master swordsman and had killed many men when he was young. One modern scholar of Chinese literature, Liu Wuji, calls him a 'banished angel . . . for his poems are unearthly, transcendental, and touched with a divine madness'.

Du Fu was of more certain heritage, born into an old and respectable (though poor) family in Hunan. His literary skills proved unaccountably inadequate to qualify him when he took the civil service exams, and as a result he attained at best only minor, badly paid administrative roles and often drifted in near-destitution. A good Confucian in his devotion to family, and often considered a model of moral integrity in comparison to the bibulous Li Bai, he was nonetheless too much a free spirit to make a good civil servant.

Both men felt the displeasure of the Tang Emperor Xuanzong's court around the middle of the eighth century, most probably because of political rivalries – it seems likely that such court infighting also undermined Du Fu's second attempt at the civil service exams in 747. They were caught up in the strife of the devastating An Shi Rebellion (755–63), when General An Lushan attempted to establish a rival dynasty in the Yellow River basin. Li Bai was accused of siding with another faction that rebelled in east-central China at the same time, and he was sentenced to exile, prompting the penniless peregrinations that engendered many of his works. Du Fu, trapped as a minor government official in rebel-occupied Chang'an in 756–7, managed to avoid similar charges, but suffered such hardships that he resigned his post and settled in Chengdu. Later he stepped down from a position there too and wandered in the middle Yangtze valley: Hubei, Hunan and the Three Gorges region.

The enthusiasm of both men (indeed, of poets generally) for wine invites comparisons with dissolute latter-day writers. The Song writer Lu You describes his own abandon: 'at the river's edge, getting drunk as I please, no regret in the world'. But whether this passion for inebriation has quite the same implication in Tang poems as it does today isn't clear. There is a sense of intoxicated rapture in it, and drinking wine incurred no particular moral judgements at that time. In this 'soaking' of the spirits, the river itself was never far away. Li Bai's most famous ode to drink, sometimes translated as 'Bring in the Wine', begins

> Do you not see the Waters of the Yellow River, coming down from Heaven,
> Rush and roll into the sea, never to return?

It's hard to separate fact from myth in the popular tale of Du Fu's death in 770. The story goes that he was swept away in the Xiang River during a rainstorm, that particular watercourse being (as we will see) a common locale for tales of drownings.* Was there genuine precognition in Du Fu's poems, in which critic David Hawkes finds an 'obsessive preoccupation with water and drowning'? Or does the prominence of those themes simply make this an apt way for the poet's death to fit the man? We can read it whichever way we please. 'The water is deep, the waves are wide', Du Fu warned – 'Don't let the water dragons get you!' In popular legend, the dragons came for Li Bai eight years before his friend: he is said to have died when he fell into the Yangtze while leaning drunkenly out of his boat to embrace the moon's reflection.

In China, drowning is a significant way to die. It was not just an inevitable danger of life on the rivers, but could be an honourable form of suicide, a thoroughly Confucian way to leave the world. Officials denied due recognition might end their lives this

---

* Another version has it that Du Fu was rescued, but died from overeating in the subsequent celebratory feast.

way in protest. When, for example, the Han official Yin Zhong, reproached by the emperor for his alleged failures after a major Yellow River flood, drowned himself around 27 BC, he was following a precedent set by the most famous drowning suicide in Chinese history.

In the third century BC the poet and statesman Qu Yuan was exiled from the state of Chu after advising the king against visiting the neighbouring state of Qin. The king went there anyway and was treacherously incarcerated until he died. Even though his warning was thus fully vindicated, Qu Yuan was blamed for the royal death. He was sent to XiaoXiang, a terrain of lakes and rivers including the Xiang and its tributary the Xiao in south-central China, below the middle Yangtze and more or less coincident with modern Hunan. In despair at the impending conquest of Chu by Qin in 278 BC, and as a protest against the injustice and corruption of the Chu court that had led to his exile, Qu Yuan strode into the waters of the Miluo River clutching a rock which bore him down under the current.

The historian Laurence Schneider says that Qu Yuan's story makes the land just south of the Yangtze 'a mythological frontier of experience, as well as a geographical frontier'. The minister's death is the founding legend of the Dragon Boat Festival that takes place on the summer solstice (its connection to criticism of state leadership tends now to be forgotten). It was said that local villagers, after taking to their boats in a vain attempt to pull Qu Yuan from the waves, beat drums and struck the water with paddles to keep demons and fish away from his body.* Implored by Qu Yuan's spirit, they threw rice wrapped in silk into the river to ward off the dragon that dwelt there. Today these parcels of steamed glutinous rice, called *zongzi*, are cooked in bamboo-leaf wrappers and eaten in the festivities, while dragon-headed boats are raced on the rivers in imitation of the frantic search for Qu Yuan's body.

The motif of drowning remains prominent in modern Chinese literature. It is by this means, not by poison or the blade, that the star-crossed lover Ming Feng meets her despairing fate in Ba Jin's

---

* One version of the legend has it that Qu Yuan's body was swallowed whole by a giant fish, which swam across Dongting Lake and up the Yangtze to the official's home town of Zigui in the Three Gorges, where the body was regurgitated for burial.

classic *Jia* (*The Family*, 1931–2)* when she is promised in marriage to an older man after being neglected by her lover. Nobel laureate Gao Xingjian's masterpiece *Lingshan* (*Soul Mountain*) is filled with tales of tragic, grief-stricken deaths beneath the waters.

Gao's meandering novel also practically drips with rain, a poetic emblem of sorrow of which Du Fu made avid use. That's not a particularly unusual metaphor, of course, but in Chinese literature it is redolent of mortal anguish. At the climax of the Yuan-dynasty play *Qiu ye wudong you* (*Rain on the Wutong Tree*) by Bo Po, the Tang Emperor Xuanzong grieves for his murdered lover Lady Yang as he hears the rain fall on the *wutong* tree during an autumn night. The image becomes a versatile symbol of the emperor's tumultuous, suicidal feelings:

> Sometimes intense like myriad pearls falling on a jade plate,
> Sometimes loud like songs and music mingled noisily at a banquet,
> Sometimes resonant like a waterfall from a cold spring at the head of the
>   blue ridge,
> Sometimes fierce like the beating of war-drums below an embroidered
>   flag.
> Alas! How the rain vexes one to death!
> How it vexes one to death,
> With its variety of sounds that stuns the ear with their hubbub!

No one living on China's great rivers could see water as an unalloyed gift of heaven. Its perils always loom, and its furies represent the overwhelming, destructive impulses of the human temperament. In 'A Second Song of Chang'an', Li Bai used a flood to depict the raging despair of a young village woman's lament after her husband has departed on the dangerous journey up the Yangtze. It's not hard to see that her thoughts are heading towards suicide by drowning:

> Last night the wind raged furiously,
> And broke the trees on the riverbank.
> The river flooded over, the waters were boundless and dark.

---

* Ba Jin, one of the Republic's most celebrated authors, originally chose the distinctly watery title *Jiliu* (*Turbulent Stream*).

This kind of poetic imagery speaks clearly enough across cultures and centuries, even if we do not see the depths it might possess to a reader immersed in ancient Chinese literary tradition. But to better appreciate how and why these artists valued and venerated water, we need to dip into its philosophical and aesthetic connotations.

The metaphysics of flow

The conjoining of philosophy, aesthetics, geography and ideas of harmony and rectitude that is evident in the *yin/yang* dualism (page 74) can also be seen in the other fragment of traditional Chinese lore that has enjoyed an often superficial popularity in the wider world: the geomantic belief of *feng shui*, literally 'wind and water'. Here is the same concordance of polar opposites, again motivated by the desire to attain harmony between invisible cosmic agencies.

Early evidence for geomancy in the pre-dynastic Neolithic cultures of the Yellow River basin can be found in astronomical alignments of built structures, a practice seen throughout the world at this time. But while Chinese astronomers in the Shang period used the sun and stars to determine cardinal directions, we shouldn't imagine that buildings were aligned with the compass in order to act as giant astronomical instruments – a plausible way to interpret some Neolithic constructions in Europe. Rather, the objective was metaphysical: by orienting structures on earth with those of the heavens, a *yin–yang* harmony could be achieved between the two.

This was one of the key motivations behind the development of the magnetic compass, which is first described in the *Mengxi bitan* (*Dream Pool Sketches*) of about 1086 by the Song polymath Shen Kuo. The compass was certainly used for travel and navigation – around 1126 the official Meng Yuanlao wrote that 'During dark or rainy days, and when the nights are overclouded, sailors rely on the compass.' But compasses were made primarily for geomancy, and some of the earliest, with the pointer made from a spoon or ladle, would never have worked on a ship.

The aim of *feng shui* was not to align buildings and structures with the fixed bearings of the heavens, but to find locations and arrangements where the flow of the pervasive energy field called *qi* was propitious. *Qi* is the source of life, and it flows where there is

A geomancer studies his compass.

harmony of *yin* and *yang*. 'Human beings are born [because of] the accumulation of *qi*', says the *Zhuangzi*. 'When it accumulates there is life. When it dissipates there is death.'

The polarities of *yin* and *yang* forged a link to the celestial bodies, but *qi* is not bound by any rigid grid: it is a dynamic entity. Fast-flowing water was particularly effective for delivering *qi*, which meant that proximity to streams and rivers was an essential criterion for the site of a building. Fed by a geomantically located source of *qi*, the building could ensure the health and success of the family who dwelt within or nearby. *Feng shui* dictated that roads, walls, paths and other structures should have winding, river-like contours that channelled the *qi* of the landscape, rather than blocking it with straight, geometrical lines.

In her introduction to a contemporary edition of *The Mustard Seed Garden Manual of Painting*, a guide written by the three brothers Wang Gai, Wang Shih and Wang Nieh in 1679, the Chinese painter and writer Mai-Mai Sze explains the significance of *qi* for artists:

> Like everything else written by Chinese painters and critics on the subject of painting, virtually the entire contents of the Manual are aimed at developing the painter's spiritual resources (*ch'i* [*qi*]) in order to express the Spirit (*Ch'i*), the Breath of the Tao . . . the constant references to the *Ch'i* are not merely rhapsodic; they are statements of a firm

conviction of the existence of the *Ch'i* . . . which was the belief underlying the whole of Chinese life in the order and harmony of nature.

In other words, the soaring mountains of the landscape painting, the
plunging streams and placid lakes crossed by frail bridges, are also
observing the demands of *feng shui*, marshalling the flows and energies
of nature to concentrate *qi*. They express a balance of tendencies:
mountains rise, rain and rivers descend. Mist exists only in relation
to mountains, a concept exquisitely expressed in the character *lan*
(岚), meaning mist, which places airy wind (*feng*, 风) below and
mountain peaks (*shan*, 山) above. Thus Chinese landscape painting
serves a similar symbolic function to the medieval cathedral inasmuch
as it offers a metaphorical representation of the order and harmony
of the cosmos. No wonder that the Six Canons (*liu fa*) of painting
from the Southern Qi period (fifth century AD) begin by specifying
that 'Circulation of the Qi produces movement of life.'

Why water is hard to paint

This sense of an animating spirit beneath superficial appearance
released Chinese painters from the obligation to attempt a faux photo-
realism (and misled some Western observers into regarding their art
as crude or unskilled). Accurate representation meant nothing if the
artist had not imbued his forms with *qi*. As the Tang scholar Zhang
Yanyuan explained:

> The representation of things necessarily consists in formal likeness,
> but likeness of form requires completion by a noble vitality. Noble
> vitality and formal likeness both originate in the definition of a concep-
> tion and derive from the use of the brush.

This need to animate water was what the Song artist and statesman
Su Shi had in mind when he praised 'lively water' in a painting, distin-
guishing it from the deadening impression of 'inert water'.*

---

* The sophisticated aesthetic of water possessed by Tang and Song scholars extended
even to the qualities needed for drinking or brewing tea. The ninth-century work
by Zhang Youxin, *Record of Waters for Boiling Tea*, reveals the depths of connoisseurship
exercised in this seemingly mundane activity. As the old Chinese proverb had it, 'The

No wonder, then, that the Wang brothers devoted many pages to explaining how one should depict water flowing between rocks. Waterfalls and torrents aren't just important elements of a picture, *The Mustard Seed Garden Manual* explains – they symbolize the very act of painting:

> Wang Wei said: 'When one is painting a waterfall, it should be so painted that there are interruptions but no breaks.' In this matter of 'interruptions but no breaks', the brush stops but the *ch'i* continues . . . It is like the divine dragon, whose body is partly hidden among the clouds but whose head and tail are naturally connected.

For this reason, flowing water should not be filled with distractions, such as the dragons and serpents, rocks and mountains that the early twelfth-century Song artist Dong You complained of in the works of lesser painters. He explained that a great artist

> causes the water to turn and twist in curves or straight lines, rippling according to its flow. Its natural long lines, fine and continuous, are orderly and without confusion. This is real water.

It was hard to get it right, and some artists acknowledged that water sources were the most difficult parts of a landscape to paint. There are those, said Rao Ziran in the fourteenth century, 'who paint a waterfall for each fold of mountain, as if they were hanging towels on a rack. This is absolutely ridiculous.' One needed always to *imply* a source of water, but not to be too literal about displaying it. As for the depiction of the more ethereal manifestations of water – mist and cloud – Chinese artists had no use for impressionistic washes or blank mirrors such as the Wangs' contemporaries in baroque-era Europe were producing. The nature of flow had to be represented with delicate tracery: 'Clouds are the ornaments of sky and earth, the embroidery of mountains and streams.' These flows are better seen not as schematic – as cartoon-like idealization – but as illustrating the channels of *qi* that surged through them. Few Western artists thought

---

quality of the tea depends on the water with which it is brewed.' In his journey along the Yangtze, the Song official Lu You comments regularly on the quality of the local waters, commending those that are 'sweet'.

Ma Yuan, *The Yellow River Breaches its Course* (Song dynasty).

about flow in anything like a comparable manner; Leonardo was an exception, doubtless because his intense interest in flow-forms was as much philosophical as it was artistic.

The ancient painters lavished great care on rivers and streams, saying that one should take five days to place water in a picture. They were an almost obligatory component of the work. 'Clouds, waterfalls and rivers are found everywhere in Chinese paintings, whether they might date to the Song dynasty or to the Qing', says art historian David Clarke. Sometimes water could be the sole subject. It was the focus of twelve studies made by the Song painter Ma Yuan; his *The Yellow River Breaches its Course* is nothing but wild waves, and all the more terrifying for that.

Water is not at all insubstantial, the Wangs explain, echoing Lao Tzu in pointing out that it can strike down mountains and pierce rocks. Water *creates* structure; more, it is the nourishing ingredient of all that exists:

> Is not water, whether trickling, flowing, spraying, foaming, splashing, or in rivers or oceans, the very blood and marrow of Heaven and Earth?

For this reason, we mustn't see the Chinese landscape as a formation of solid earth and rock that water crosses superficially. Rather, the mountains and water are intimately entwined: *yin* and *yang*, both of them ingredients integral to the whole composition. That is why traditional Chinese landscape painting is simply called *shanshui*. It was almost a sexualized union, certainly a fecund one: it was thought that water is produced when clouds touch the rocks of mountaintops, which were known as 'cloud roots' (*yun gen*). The early Qing painter Shitao, a Daoist and a contemporary and friend of the Wang brothers, claimed that

> The immensity of the world is revealed only by the function of water, and water encircles and embraces it through the pressure of mountains. If the mountains and water do not come together and function, there will be nothing to circulate with or about, nothing to embrace. And if there is no circulation and embracing, there will be no means of life and growth.

Shitao explained that water has in fact a pictorial solidity equivalent to that of stone:

> I know from my own perceptions that mountains are oceans and oceans are mountains; and they seem to understand that I have perceived this about them. It is all conveyed by the artist through his flowing movements of brush and ink.

This equivalence of *shan* and *shui* is plain to see in several of Shitao's works, such as *Landscapes Depicting a Poem of Huang Yanli* (1701–2) and *The Force of Rivers* (1703), in which waves become fields of hillocks.

It's no coincidence that water is the primary medium of the Chinese artist, in contrast to the egg tempera and oils of the West. This is more than just saying that a watery wash is good for representing water itself – to the Chinese artist, the medium was the message, since only with so fluid a substance could one capture the vitality, the qi, of what was being depicted. This almost alchemical communication of a spiritual quality through the qualities of the fluid ink (*shui mo*) was made explicit in the inscription Shitao added to a landscape painting of 1702, where he explains what the brushstrokes should accomplish:

> First, the transformations of water; second, the introduction of these into the ink; third, their reception by [the ink's] potentiality.

This is not so different from the union of body, spirit and medium that the modern Abstract Expressionists sought, which is why there is more than casual visual analogy in the idea that Shitao anticipated the dynamic vitality of Jackson Pollock. He advised that the movement of the painter's wrist should be 'flowing deep down like water or shooting upwards like flames' – like water, it should 'move naturally, without the slightest bit of coercion'. If one's wrist is sufficiently responsive, Shitao asserted, a single stroke could show the entire universe.

And so, 250 years before Pollock splashed canvases with seemingly spontaneous and random flicks of his dripping brush, we find Shitao covering a landscape with '10,000 ugly inkblots', dissolving the picture into a chaotic mass of blotches that assert a kind of wild beauty through their very ugliness. The Tang painter known as Wang Mo went even further, his antics reminiscent of those of a contemporary performance artist, as the ninth-century book *Tangchao minghua lu* (*Celebrated Tang Painters*) relates:

> Whenever he wanted to paint, he would get himself half drunk first, then dance around splashing the ink frantically onto the silk, spreading it out freely with his hands or sweeping it with a brush to obtain different intensities. The shapes of the ink splashes were to suggest what objects they were going to be: a mountain or a piece of rock, a stretch of water or a piece of cloud. Every movement was free and improvised. The result was fascinating and highly dramatic.

Sometimes Wang was said to have used his own hair as a brush. It's easy to see how he got his nickname *mo*, meaning 'ink'.

All this is about as far from pictorial realism as you can get, and reinforces the argument that the artists of China who seemed so naive and primitive to early Western emissaries were in fact already embodying, in an exquisitely codified and indeed virtually conservative manner,

Shitao, *10,000 Ugly Inkblots* (early eighteenth century).

the now very modern idea that art is about expression more than it is about representation. To the untrained eye, Chinese art can seem repetitive, even hackneyed: *more* rocks and rivers, yet another poet gazing on a stream. Even the scenes are often generic, an assemblage of trees, waterfalls and peaks that might look little like the alleged location. But it was all about style. A painting was judged not by how elegantly or accurately the artist had depicted the subject matter but by what the brushstrokes revealed of the artist's moral character. The use of optical instruments like the camera obscura by Western artists such as Johannes Vermeer to enhance the veracity of their works would have seemed to their Chinese contemporaries to be utterly incomprehensible, or perhaps, rather, simply nothing to do with art.

A landscape, then, needed the artist to give it meaning, not just to capture it in faithful rendition. The artist wasn't aiming to represent *shanshui* so much as to transform it: only through the consciousness of a poet did nature become 'scenery' (*jing*). The *qi* of the universe acted through the *qi* of the painter, and what resulted was a timeless portrait of harmony. Time passes as the flowing of water – but as Confucius said, that flow is itself eternal. And while Western art relied on narrative – particular people and events, firmly located in time – the figures in Chinese landscape painting are often anonymous and insignificant. Even to the untrained eye the implication is clear: the human world is insignificant amid the serene splendours of the universe. We are all just part of the immense cosmos, out for a stroll.

Currents of dissent

The Tang poet Wang Wei tends to get eclipsed by his two more famous contemporaries Li Bai and Du Fu. But he exemplifies rather better the polymathic nature of the intellectuals of that era, being renowned as much for his painting and music as for his writing. His poems convey pure appreciation of nature for its own sake, as in 'Bamboo Lodge':

> I sit alone in the dark bamboo grove
> Playing the zither and whistling long.
> In this deep wood no one would know –
> Only the bright moon comes to shine.

Liu Wuji argues that Wang's abilities as a painter merged with his skill at poetry: 'Just as a painter creates his landscape with a few strokes of the brush, so Wang Wei, the poet-painter, produced in his poems poetic pictures with a few choice expressions.'

Yet artists like Wang Wei were not aloof, ascetic nature-worshippers. They were worldly men, officials engaged in affairs of state. Painting as a profession was rare; one painted, just as one composed poetry, as an extension of the self-cultivation that every educated person was expected to practise. From the Tang to the Qing dynasties, artists were often also civil servants: so-called literati who had demonstrated their intellectual grasp of the philosophical classics in the imperial examinations (*jinshi*). Although these were a small, usually wealthy elite (for education cost money), their artistic endeavours were not so much a leisure pursuit or a specialism dependent on innate talent, but a skill expected of all gentlemen.* These works showcased the artist's subtlety and sophistication of thought, and by alluding to the canon of classic writings they might foster confidence in his scholarship and seriousness of mind. In short, your paintings and poems revealed what sort of person you were. The literati interjected their own thoughts and ideas in their work, making 'self-fashioning' an aspect of Chinese art many centuries before that tendency emerged in the West. Household guests might be asked to mark an occasion with a poem; friends composed verses for one another when they parted. Yet for

---

* A tradition that regards skill at composing poetry as a measure of personal virtue and intellect persists in contemporary China. Political leaders have written poetry – dutifully and admiringly quoted by their subordinates – while skill at completing couplets known as *duilian* has long been used to test the credentials of suitors for a daughter.

The Tang poet Wang Wei plays the zither and contemplates next to the river at Bamboo Lodge in Wangquan, as depicted in the Ming-era *Tangshi huapu*. Poets frustrated by failure as officials were drawn to such places: as the Song writer Ouyang Xiu wrote in a preface to a collection of poems, 'Scholars who cherish great thoughts but fail to apply them in life are often fond of visiting mountain peaks and riverbanks.'

all the personal investment, these works were often produced with an affectation of insouciance. Shitao liked to intimate that he was a mere dilettante with the paintbrush, words being his real forte.

For some officials, composing poems and paintings did not much more than fill the hours when their posts demanded little from them, and certainly a great deal of mediocre art was produced. But others were highly gifted artists. And as political players, they had things to say, perhaps of a controversial or critical nature. It is precisely because Chinese landscape art was referring to something beyond itself that it could encode messages in stone and stream that the cognoscenti would understand. In this way, criticism that might otherwise land an official in trouble could be kept discreet and ambiguous – for Chinese rulers were rarely tolerant of open challenge. Political dissent was not necessarily directed at an emperor in person, for many government policies were determined, or at least shaped, by ministers, and the imperial courts were inclined towards factionalism. When an artist was accused of planting treasonous meanings in his work, then, the real reason for the charges might be political infighting and the acting-out of power games.

How, though, could a political meaning be expressed in mountains and water? Take the 'Eight Views' format that East Asian landscape painters have used for many centuries, examples of which can be found in the traditional art of China, Japan and Korea. These bucolic scenes testify to a profound sensitivity to nature. The composition consists of a series of eight views of a region renowned for its natural beauty, the most prominent elements being tall mountains, lakes, rivers and mist. *Mountain Market: Clear with Rising Mist* by the thirteenth-century Song painter Xia Gui is a typical example. The market is nothing but a handful of little roofs, and it is far from bustling: two solitary figures move between the buildings, sheltered by pines. This remote outpost of human enterprise is overshadowed by soaring peaks, plunging down into a valley lost in mist that rises from the river winding between the hills in the foreground.

The 'Eight Views' tradition is generally thought to have been started by Song Di, a minor Song official in the eleventh century. Song Di's seminal *Eight Views of XiaoXiang* no longer survive, but from the renditions – more homage than copy – of those artists who followed him we can deduce that they shared the delicate poise that we find in the landscapes of Xia Gui. All the same, there was nothing polite and serene about Song Di's paintings. They were, on the contrary, works expressing political disillusion, and they came close to getting him into serious trouble. You could, as we'll see, say a great deal with mountains and water – because in China, water is always a political issue.

Artists of the Song time had no illusions about what it might mean for their work to provoke the displeasure of their rulers. They only needed to consider the example of Song Di's friend, the artist and poet Su Shi, who fell foul of court conflicts. A promising young government official under the Song Emperor Renzong, Su Shi's fortunes took a turn for the worse when the emperor died in 1063 and his successor, Shenzong, favoured the fiscal 'New Policies' of the

Mountain Market: Clear with Rising Mist by Xia Gui (active *c*.1195–1230), one of the painter's 'Eight Views'.

statesman Wang Anshi, a reformer who belonged to a rival political faction. On that occasion, Su Shi suffered nothing worse than being removed from the court circle in Kaifeng and sent south to govern Hangzhou. And after Wang Anshi's policies failed, possibly as a result of a drought in 1074, for a while Su Shi was rehabilitated.

Yet the political battles continued, and Su Shi's enemies were able to argue that his poems attacking the New Policies were seditious. He was recalled to the capital of Kaifeng, imprisoned and put on trial. Su Shi explained that his target in the verses had been the officials responsible for the policies, not the emperor himself, and that he was only observing the Confucian obligation to speak out when the emperor was receiving bad advice. It did him no good. In what became known as the Crow Terrace Poetry case (the Office of the Censor at the imperial court was known as Crow Terrace), in 1079 he was sentenced to death by beheading. It was pointed out, however, that the first Song Emperor, Taizu, had forbidden the execution of officials, and so Su Shi found himself instead exiled to a remote village on the Yangtze. He was eventually sent still further away to the fringes of the empire on disease-infested Hainan Island, where he died in 1101.

The incident led poets to be more cautious. Su Shi himself continued to express his views in verse, but he buried them more deeply so that only a few knowledgeable individuals would spot them. The allusions became extremely subtle: a poem might, for example, use the same rhymes as another one that was itself more outspoken. To catch these meanings, you needed a thorough familiarity with the canon. And were anyone to raise accusations, the poet could present another, more innocuous interpretation of his words.

This cryptic art of dissent was adopted by painters too. The danger of painting explicitly political subjects became clear in 1074, when the artist Zheng Xia depicted refugees from the terrible drought that may have derailed Wang Anshi's schemes – a tradition that became known as *liumin tu*, 'images of refugees'. To show the common people in such an abject state was considered an imputation against the governance of Emperor Shenzong, and Zheng Xia was banished.

In that same year Song Di fell foul of court politics. He was dismissed from his post, officially because of his culpability in a fire that had swept through the State Finance Commission and destroyed many important documents. The fire started from an unattended stove in the buildings of the Salt Monopoly, of which Song Di had earlier been the head. Since he had left that position three months before the fire occurred, however, it seems most likely that the incident was a ruse by Song Di's enemies to discredit him. He was dismissed without a stipend.

That injustice was what seems to have motivated Song Di's *Eight Views of XiaoXiang*. The titles alone* signify that the works were all concerned with water-related imagery:

Geese Descending to Level Sand
Sail Returning from Distant Shore

---

* These titles were appended to the works by others; Song Di did not name them. Some connoisseurs considered it unseemly to give paintings titles. 'Nowadays,' wrote the Song intellectual Liu Xueqi haughtily, 'those painters who fix titles to their works before [painting them] are not scholars.'

Mountain Market, Clearing Mist
River and Sky, Evening Snow
Autumn Moon over Dongting
Night Rain on XiaoXiang
Evening Bell from Mist-Shrouded Temple
Fishing Village in Evening Glow

They seemed to be images of arcadian serenity, the sort of paintings that *The Mustard Seed Garden Manual* would later commend. But what were they really saying?

It surely wasn't just for its visual appeal that Song Di had selected XiaoXiang as his setting. In the Song era this was considered a backward and barbaric region, and was consequently a common place of exile. It was to here that Qu Yuan, the archetypal honest Confucian official unjustly spurned by his rulers, had been banished in the Warring States period – and where, as we saw, he drowned himself in protest, forever associating XiaoXiang with critique of the state.

The legendary Emperor Shun was also said to have died in XiaoXiang while on an inspection tour. His several wives travelled there to pay their respects, but were so overcome with grief that they threw themselves en masse into a river, perhaps the Xiang. XiaoXiang had thus become linked to the idea that rulers as moral and virtuous as Shun were hard to find. It was said that the spirits of the place could distinguish a good ruler from a tyrant. When the despotic first Qin Emperor sailed down the Yangtze to XiaoXiang in search of a shrine that had been erected to Shun's faithful wives, his entourage was held back by a great gale that endangered the boats, as though the elements themselves were protesting the emperor's bad faith. Enraged by nature's effrontery, he had 3,000 convicts cut down all the trees around the shrine and paint the mountain red.

Aside from the choice of location, there were other ways that a Song painter could imbue a *shanshui* work with political overtones. At the beginning of the Song period, landscape was typically painted in the 'high-distance' style in which mountains soar to lofty heights, dwarfing figures or dwellings situated around tranquil waters in the foreground. This composition was considered to celebrate the stability of a powerful and hierarchical state. 'A great mountain', wrote the scholar Guo Xi in the eleventh century, is dominating as chief over the assembled hills, thereby ranking in an ordered arrangement the ridges and peaks, forests and valleys as suzerains of varying degrees and distances. The general appearance is of a great lord glorious on his throne and a hundred princes hastening to pay him court.

But painters of this period began increasingly to use instead a so-called 'level-distance' perspective in which the peaks were lower and more distant and the composition was dominated by wide rivers and lakes: a vision less overshadowed by the oppressive presence of court authority and protocol, and which might symbolize a removal from that stifling milieu because of expulsion and exile. According to Alfreda Murck, a specialist on Song artistic culture, 'The growth in popularity of the level-distance compositional mode among scholar-officials in the late eleventh century can plausibly be ascribed to the large number of officials whose careers ended early due to factional politics.'

This subtle articulation of dissent was a function long served by poetry too. Du Fu was the role model for the Confucian artist-official who, after being spurned by the government, used his work to criticize the authorities while remaining loyal to the nation. (Whether Du Fu himself fitted that role is debatable.) In the poems Su Shi wrote to accompany Song Di's *Eight Views*, he made many allusions to the works of Du Fu, which the well-informed reader would understand to imply parallels between the great Tang poet's unjust treatment and the grievances that both Su Shi and Song Di had suffered.

One must wonder, however, whether these allusions were sometimes so obscure that they defeated the object. It's said that a connoisseur might need to spend many hours contemplating a painting or a poem, carefully cross-checking its references, before the meaning would become clear. There are tales of literati who would cry out in joy as the true implications dawned on them. Artists who courted the emperor's favour, on the other hand, would kowtow to the officially endorsed modes of expression. Their art would dwell on the optimistic and auspicious – the blooming of a rose, say – and it would be painted with meticulous realism and a suppression of all ego and personal expression. The Song Emperor Huizong, just eighteen years old when he came to power in 1100, was a great patron of the arts and was himself a gifted painter and poet (some say at the expense of his official duties), but he had strict ideas about matters of content and style, and his entourage of court painters was carefully vetted. Nothing was more to his liking than the much-imitated masterpiece of Zhang Zeduan from the early twelfth century, *Along the River During the Qingming Festival* (see page 122), which unscrolls into a panoramic view of life in the capital Kaifeng and the surrounding countryside, united by the river that passes through them. Here is a vision of a prosperous and well-ordered world: an ideal society rendered harmonious by the virtue and wisdom of its governors.

Flood as critique

The painter Shitao was still a small child when the Ming dynasty crumbled in 1644 before the advance of the Jurchen invaders from Manchuria. The Jurchen tribes were unified in the early seventeenth century by the chieftain Nurhaci, and when his son and successor, Hong Taiji, died in 1643, Hong's ninth son, Fulin, was made the third ruler of the Manchurian state, which Hong had awarded the dynastic name Qing. Fulin was just five years old at the time, and state affairs were dictated by two Manchu co-regents: Fulin's uncle Dorgon (Nurhaci's fourteenth son) and Nurhaci's nephew Jirgalang.

The Jurchen were initially welcomed into China as allies of the Ming general Wu Sangui, whom they helped to repel the usurper Li Zicheng. But once this task was completed, Dorgon's troops made it plain that Wu's forces were now subservient to his wishes. They entered Beijing unopposed and, claiming to be 'avenging' the deposed Ming Chongzhen Emperor, assumed the Mandate of Heaven. Fulin thus became the first Qing Emperor, the Shunzhi Emperor.

Like the Mongol invaders who founded the Yuan dynasty in the thirteenth century, the Manchurians assimilated many aspects of Chinese culture. One exception, however – a source of deep humiliation to the Chinese – was Dorgon's insistence that all male citizens should adopt the Manchurian shaven head and long pigtail as a sign of allegiance to the Qing state. Refusal to comply carried the death penalty. This style, characteristic of the Chinese 'coolie' to Western eyes until the end of the Qing dynasty, was considered unmanly by the Han Chinese, and its imposition implanted long-standing resentment towards their new rulers.

Shitao, *The Crosses Torrent* (1699).

Yet by the early eighteenth century the Qing empire seemed to many outsiders to be a model of stable governance. The Kangxi Emperor (1661–1722) was the first of the 'big three' (Yongzheng and Qianlong followed) during the golden age of the Qing. Shitao, like many other Chinese, was able to accommodate himself to the reign of the northern invaders without ever becoming entirely comfortable within it. He felt his heritage was still that of Ming culture, and his landscape paintings display a preoccupation with the disruption and dislocation of dynastic change. Commissioned to paint for the Kangxi Emperor or for high-ranking Qing officials, he too used art as a medium for understated criticism of state policy. The waterways and their (mis) management provided a perfect subject for this.

Some of Shitao's watery landscapes have a stately elegance to them: scholars boating in tranquillity, the vast expanse of the Yangtze delta, morning mist in the mountains. But no one was better at depicting ominous storms and torrents: waters 'boundless and dark', as Li Bai had it, expressing a profound disquiet. In *The Crosses Torrent*, raging waves threaten to drown a peasant trying to cross a river on his ox while the winds whip through the trees. In *The Sound of Thunder in the Distance*, a man seems to sense disaster looming in the heavens.

Shitao occasionally risked making the political connotations more explicit. In *Desolate Autumn in Huai-Yang* (1705) he painted the devastating flood which occurred that year around Yangzhou: the kind of natural disaster that always raised questions about the mandate of the rulers. This flood was particularly compromising for the Kangxi Emperor, since he had visited the region in the summer and inspected the dykes at Qingkou, where the Huai River intersected the Grand Canal. There had been bad flooding here in 1699, but the emperor had nevertheless been optimistic about the situation six years on. 'The water has gradually been returned to grain transport use', he announced:

> The banks were raised higher than the water, and now they are over a *zhang* [about 3.5 metres] higher . . . From the look of things, our hydraulic project is completed – our heart is truly joyful.

But only three days after the emperor returned to Beijing, the rains in Jiangsu began to fall. By early August, Shitao wrote,

> there were vast waves for a thousand *li* all through Huai [the area of Huai'an] and Yang [Yangzhou]. The two prefectures together formed a marshy kingdom; faced with this frightening sight it was hard not to be anxious.

And rightly so. At the end of August the dykes broke and the waters of the Huai gushed forth. In *Desolate Autumn* the city of Yangzhou is confined in the corner of a great expanse of water, overwhelmed by nature. In the long inscription to the work, Shitao refers to the crumbling of the Sui dynasty through neglect and degeneration, whereby it had forfeited the Mandate of Heaven:

> Heaven loves to see people born, but people do not assume their responsibility;
> When humans depend on worldly desires, Heaven does not take responsibility for them.
> When the relation between Heaven and humans reaches this point, how can one not be concerned?

At first glance this might look like a naked accusation that the Qing Emperor had neglected his duties of water management, and in consequence had brought into doubt his right to govern. But the intended implication was rather different. There was much public concern at that time about the dissolute habits of the Kangxi Emperor's son and heir apparent, Yinreng. When the emperor had visited Nanjing and had been inadvertently given a sitting mat with grubs infesting it, his son had called for the popular prefect of Nanjing to be executed. The unfortunate prefect narrowly avoided that fate, but people saw the incident as an example of the prince's arrogance and brutality. Yinreng was also notorious for demanding that local salt magnates bring him young boys and girls for his pleasure. Against this context, it seems clear that in his flood painting and poem Shitao was not criticizing the emperor for neglect, but was exercising his Confucian obligation as a government official to warn his ruler that this archetypal 'bad son' threatened to undermine his authority.

Such respectful criticism was not only tolerated; it was heeded. After further scandals, the emperor stripped Yinreng of his status. Here, then, flood imagery supplied a multilayered code, rooted in hydraulic history, through which the artist could offer an acceptable critique of the imperial court. That Shitao was not, after all, a revolutionary was clear from the painting he made in 1689: *The Seas are at Peace and the Rivers are Pure* echoed the old notion that, when the ruler was just and wise, heaven smiled and the waters were calm.

Culture and revolution

If Shitao's water paintings sometimes showed a desolate world beneath angry heavens, those of the twentieth-century artist Fu Baoshi (1904–65) are positively saturated with foreboding rain, sometimes almost obliterating the subject matter and merging with rivers and lakes.* Works such as *Spring Rain at the West Lake* (1963) create, in David Clarke's words, 'a world on the verge of dissolution'.

---

* Fu Baoshi's apparent fascination with water, and especially rain, finds both a pleasing resonance with his father's occupation as an umbrella repairer, and a delightful irony in the fact that he apparently had a phobia of water and couldn't bear to immerse himself in more than a few inches of it.

This has earned Fu a reputation as the 'Chinese Turner'. But J. M. W. Turner's impressionistic watercolour washes were never so politically charged, for a world on the verge of dissolution was hardly how China's modern leaders wished the country to be seen. Yet Fu astutely navigated the stormy currents of his country's modern era. In the 1920s and 30s he aligned with the Nationalists, but in the People's Republic he became (after a spell of 're-education') a cultural official who advocated the use of traditional methods in art to represent approved political and social messages. At face value he seemed perfectly content with Mao's insistence, in a talk in 1943, that all art should advance the cause of revolution.

It's hard to know if Fu was really endorsing the party line, or instead returning to the old tradition of using water imagery for subtle, coded criticism of the state. Understandably enough, he made no explicit statements that would enable us to decide. Take, for example, the way he painted Mao's swim across the Yangtze. While several modern Chinese artists chose to celebrate this feat in crude heroic style, Fu's version is difficult to read: Mao appears almost overwhelmed by the waves, not so much waving as drowning.* And when he painted the famous suicide of Qu Yuan, was Fu implying that today officials needed to take a stand against political corruption and repression? His 1961 painting inspired by Mao's poem 'Heavy Rain Falls on Youyan' echoes Mao's message about the destructive force of water but doesn't trouble itself with the final stanza in which Mao suggests that Communism has finally gained the upper hand over nature. Here Fu seems to take the side of water, not of the state.

What, too, should we make of Fu's *Heaven and Earth Glowing Red* (1964), where the wind and rain threaten to overwhelm the red sun, an obvious symbol of the Communist state? But these readings are ambiguous, and in the absence of any commentary from the artist we are playing a guessing game. Was he celebrating the state, or offering a sly challenge? Whatever the case, he doesn't seem to have been judged

---

* Official images of this feat are no more reliable a record. A much-reproduced photo of Mao's 1966 reprise, in which the Chairman is surrounded by security guards, is faked with almost comical transparency: the five geometrically arranged heads display impossible tonal contrasts, there is not a hint of the bodies to which they are supposedly attached, and they are devoid of any sheen of water. To judge from this image, the Great Helmsman's swimming style was stately to the point of being statuesque.

adversely; his optimistic vision of a red dawn, *Red Sun Rising* (1962), painted with Guan Shanyue, was rechristened by Mao *This Land So Rich in Beauty* (Mao wrote the inscription himself on the massive work) and is given pride of place in the Great Hall of the People in Tiananmen Square. But perhaps it was Fu's good fortune to have died in 1965, just before the harsher adjudication of the Cultural Revolution.

Others were less fortunate. In the early 1950s artists were obliged to make little more than token gestures to state ideology, for example by adding a red flag to their landscape. But towards the start of the following decade, that ideology was increasingly expected to predominate in all aspects of culture and society. Mao's wife, Jiang Qing, for whose artistic sensibilities 'vulgar' and 'philistine' would be generous descriptors, proclaimed that all art should be *'hong, guang, liang'* (red, bright and shining), and she organized 'black painting exhibitions' equivalent to the Nazi displays of 'degenerate art'. Older artists were condemned as feudal relics, and their work would be criticized if not actively destroyed by Red Guards. Ink painting smacked of the discredited 'Four Olds'. The official preference was for a realist style using oils, and in a revealing echo of the Western medieval tradition

Fu Baoshi, *After Mao Zedong's Poem 'Swimming'* (1958).

for religious art, only a chosen few (young Red Guards) were entrusted with painting the head of Mao.

What the state now demanded were not mournful visions of rain like those rhapsodized by Du Fu centuries earlier, but optimistic images that reflected the bright promise of the Communist state. Even a cheerfully realist work like *Spring Rain* (1963) by Huang Wenbo was considered suspect because it showed sheltering peasants apparently at the mercy of nature's water rather than benefitting from the state's ability to control it. What the authorities wanted to see were images like Li Hua's *Conquering the Yellow River* (1959), with the people working together in harmony to harness water through engineering on an immense scale, or Wei Zixi's *The Yangtze Becomes a Thoroughfare* (1973), celebrating Mao's dream of a bridge across the Long River. Zheng Shengtian's *Man's Whole World is Mutable: Seas Become Mulberry Fields* (1968) shows Mao inspecting the Yangtze valley in a messianic pose: a modern water god. (Zheng, who was denounced and disgraced for voicing concerns about the violence of the Cultural Revolution, was not permitted to paint Mao himself – that task was given to the more ideologically sound young artists Zhou Ruiwen and Xu Junxuan.)

Li Hua, *Conquering the Yellow River* (1959).

Some traditional artists tried to adapt to the new expectations with a shotgun marriage of new and old: Ying Yeping's *The Lofty Mountain Bows its Head; the River Yields to a Road* (1956) has Ming-style *shanshui* in the background yielding to a construction project in the foreground, as nature is tamed by socialism and the landscape literally cut down to size. It is a shockingly ugly collision.

Performing in water

Contemporary Chinese artists who have used water as a medium for guardedly oblique references to state repression and corruption can thus tap into a very old tradition with a well-developed vocabulary and set of allusions. According to David Clarke, 'much water-themed contemporary art in China can be best understood as a contestation of state rhetoric concerning the control of water' – and, by extension, control of life.

One would have to be a fairly dull-witted official not to recognize that, by attempting to imprint the character *shui* (水) on the Lhasa River in Tibet using a gigantic carved seal, the artist Song Dong in his performance piece *Printing on Water* (1996) is commenting on the futility of efforts to impose Chinese rule and Han culture on the roof of the world. And when in 1994 Wang Jin staged a performance called *Battling the Flood* in which he dropped fifty kilograms of a red pigment into the Red Flag Canal on the border of Hebei and Henan – a flagship irrigation project of the Great Leap Forward – the livid current became a challenge to the slavish 'red' rhetoric of the Mao era, although it can equally be read as a symbol of the blood spilled during those turbulent times – or a comment on today's river pollution.

Song Dong's 1995 work *Writing Diary with Water*, in which he records his text with brush marks on a stone slab using water alone, can be read (almost literally) as an exploration of Buddhist impermanence – although in fact Chinese calligraphy using water is a well-established way of practising brush skills without wasting expensive ink and paper. But in Zhu Qingsheng's performance piece *Tonggu Character Style* (1997), in which the artist wrote poems in ink on silk scrolls immersed in a stream running through Huairou village in Beijing, the work carries heavy implications of censorship.

The close and ancient links between brushstrokes, characters, their meanings and their fluid medium of ink offer artists a rich set of metaphors from which to weave subtle messages. In Dai Guangyu's performance piece *Demonstration of Shui Mo Painting Skills*, where flowers were 'washed' with water using the traditional brushstrokes of the calligrapher, it's not obvious that the avant-garde artist intended any political undertones. They're more apparent in his *Landscape, Ink, Ice* works, where the characters *shan* (山) and *shui* (水) drawn in broad ink strokes on the surface of a frozen lake comment on the fragility and evanescence of traditional Chinese culture. There is no mistaking the political criticism in Zhang Peili's video work *Water – Standard Version from the Dictionary Ci Hai (Sea of Words)* (1991), in which a well-known television news presenter reads out characters containing the water radical and other water-themed terms: the meaninglessness of the message undermines her authority as a state-approved voice. Why water? Clarke argues that 'there must have been a sense on Zhang's part that water's innate nature allows it to easily play the role of symbolic opponent to rigidity'.

For traditional painters too, water-related subjects continue to provide a time-honoured lexicon of discreet dissent. Liu Xiaodong's multi-panel, fragmented oil paintings of the displaced populations and new settlers at the Three Gorges Dam have echoes of the heroic realist style of the Communist era, but could hardly be interpreted as a celebration of the project. These people, says Liu, 'don't know what their fate will be, they don't even know what's going on, they just find themselves in the scene'. That, indeed, might be an epigraph for the New Realist artists of which Liu is one of the best-known representatives, who place themselves outside of both state propagandist art and avant-garde reaction to it: they simply observe everyday life, often in a highly realist manner using oil techniques, and leave the observer to draw conclusions.

The Three Gorges Dam has become a particularly strong locus for politically oriented art, albeit often with an ambivalent flavour that resists categorization as simple protest. These works often evince a need to mark the changes, but not necessarily with the intention of expressing regret or outrage. In 1995, shortly after construction began, the artist Zhuang Hui went to three famous locations in the Three Gorges – at Sandouping, Wu Gorge and Baidicheng – and drilled holes with an ancient instrument called a Luoyang shovel, allegedly invented by tomb robbers. He took photographs of the excavations, and then in 2007 sent a photographer to record exactly the same spots – where now nothing could be seen but the surface of the water. What had changed, he insisted, was not the holes in the earth, but the river itself: 'Although it's still called the Yangzi River, it's not the same.'

There is even more ambivalence in the Three Gorges project of Chen Qiulin, who grew up in the city of Wanxian that was half submerged by the reservoir. In a video of 2002 called *Bie fu* (*Rhapsody on Farewell*) she documented bulldozers levelling buildings, disoriented people leaving on boats with their bundled belongings, and homeless displaced citizens in temporary huts and shelters. The tone was elegiac and disenchanted. 'It felt like these things were being snatched away and there was nothing I could do about it', she said.

But when she returned to the site for her 2006 video *Color Lines*, she seemed less pessimistic about the changes. The ruins were gone, and in their place were modern buildings and placid waters. Chen appears herself in the film as an angel dressed in fabric used on the construction site, and she seems to be as much heralding the new world as marking what has been lost. That cautious optimism persisted in Chen's subsequent film *The Garden* (2007), in which she hired migrant workers to place big ceramic vases of peonies throughout the new city among residents going about their daily work.

Liu Xiaodong, series from the *Three Gorges Project* (2003–6).

Others have been less sanguine. Ji Yun-Fei, an artist from Chengdu, used traditional brush-and-ink techniques in a modern style to paint a scroll called *Water Rising* (2006) on which he showed displaced people moving their possessions in bundles and on carts. Because of both the subject matter and the methods, Ji's work makes an explicit link with the tradition of *liumin tu* refugee scenes that got Zheng Xia into such trouble during the Song dynasty (see page 274).

In this way, artists like Ji have been able to imply connections between the rising waters of the Three Gorges project and those of the ancient floods, which also turned people into homeless refugees – and which were often blamed on official incompetence and neglect. In *Below the 143 Meter Watermark* (2006), Ji again mobilized tradition to make his point, depicting a classic *shanshui* landscape that would entirely vanish once the reservoir waters rose to their highest level. But even Ji, who in 2001 visited the largely abandoned Yangtze cities before they were flooded, is equivocal about the massive hydro-engineering project: 'It is not simplistically right or wrong', he said. 'The issue is complex.' He regards the loss of the traditional Three Gorges landscape almost as a symbol of China's more or less system-atic destruction of its past during the Cultural Revolution. 'There are two periods of great cultural destruction in Chinese history', he said in 2004, when the water level began to rise. 'The first was the Cultural

Revolution, and the second is happening now.' Every child in China knows all about the legends of the Three Gorges, he says, but now this area is to be changed forever. 'It physically represents what we have done for so many years.'

In Zhong Ming's 1996 film *Wushan yunyu* (*Rainclouds over Wushan*), the 'rainclouds' are also the imminent floodwaters of the Three Gorges Dam, which will drown much of Wushan. The title alludes both to Mao's reference to a dam that will 'hold back Wushan's clouds and rain' in his poem 'Swimming', and to a traditional euphemism whereby *yunyu* stands for sexual intercourse. The wall of concrete that suppresses the river's flow, and the floodwaters that rise behind it, therefore become erotically charged.

The depiction of people in water is freighted with political resonance, since such images almost inevitably comment on Mao's swimming exploits – themselves arguably acts of carefully orchestrated political performance art. Zhuang Huan's photograph of migrant workers paid to stand passively in a pond in Beijing – as the title explains, *To Raise the Water Level in a Fishpond* (1997) – not only lampoons the pointless water-engineering feats of the Mao era and beyond but also contrasts these passive figures with Mao's vigorous swimming. It's doubtful, however, that any artistic comment on that event has been as direct and shocking as that by Liu Wei, called *Swimmers '94* (1994), which shows Mao bathing, tongue lolling lewdly, in the company of a brash naked woman with legs splayed – a comment not so much on Mao himself but on the vulgar fate of his socialist state today, in thrall to an exhibitionist consumerism.

Likewise, Wang Ziwei pretty much throws caution to the wind in his pop-art series *Mao Wave Dance*, begun in 2000, in which multiple cartoon Maos skip with kitsch effeminacy among traditionally styled breaking waves. Even as late as 2006, the Chinese authorities demanded the removal from a Beijing gallery of a painting by Gao Qiang of Mao, his skin coloured a sickly yellow, swimming in blood-red water.

Documentary treatments of hydraulic projects have also mobilized water to make broader political statements. *Yanmo* (*Before the Flood*), a 2004 film by Yan Yu and Li Yifan, was fairly literal in its depictions of the corruption and tensions surrounding relocations from the

Liu Wei, *Swimmers '94.*

city of Fengjie in the Three Gorges – although scenes of individual profiteering and cheating belie any conclusion that the locals are passive, innocent victims of official misdeeds. In a series called *River Elegy*, broadcast on Chinese television in 1988, the muddy waters of the Yellow River were contrasted with the clear blue sea as if to juxtapose the stifling, destructive and inward-looking Mao era with the freedom, prosperity and internationalism promised from unseen foreign shores. *River Elegy* captured the sense of optimism and impending change that, a year later, was harshly crushed in Tiananmen Square; after the crackdown, the documentary was banned and its director, Su Xiaokang, had to flee the country.

Yet amidst the savage satire and angry protest of these works, drawing on the rich cultural symbolism of water in China, there are moments of poignancy in which artists reflect on what, in the modern country, the once revered and celebrated waterways have become. In *Washing the River* (1995), Yin Xiuzhen turned some of the polluted waters of the Funan River – the 'mother river' of Sichuan, which flows through Chengdu – into blocks of ice that were stacked on the river's banks. She invited those passing by to wash the ice with fresh water: for every culture recognizes that water is the universal cleanser, but what will clean water itself?

Such artistic statements should trouble China's leaders more than irreverent depictions of Mao amidst the waves, because they address an issue on which China's future depends: what has become of its water today?

10 Water and China's Future

Threats, Promises and a New Dialogue

> Thus we find, in this country [China], the strangest mixture of
> wisdom and folly.
>
> Nicolas Boulanger, *The Origin and Progress of Despotism*
> *In the Oriental and Other Empires* (1774)

There are two big problems with water in China today. There is not
enough of it to go round, and it is often so foul that no one can use
it anyway.

Both of these are new problems. In the past, China struggled to
overcome the dominance of water's might and excess. Sometimes, it
is true, heaven withheld it so that crops withered and the ground
cracked. And some old water and land management practices damaged
the environment. But the difference today is that, on top of whatever
challenges nature might pose, human activity has made the problems
far worse. Population growth and the intensification of industry, agriculture and urbanization have placed impossible demands on water
resources, while at the same time contributing to their almost ubiquitous degradation. China's water problems are now interconnected:
pollution, damming, overuse, land reclamation and climate change
combine to devastating effect.

These are political problems as much as they are social, economic
and ecological. As the political scientist Judith Shapiro observes in her
searing and penetrating book *Mao's War on Nature*, 'the abuse of people
and abuse of nature are often interrelated'. That tends to be true

anywhere, but the truth goes deeper for China's water than it does for other aspects of environmental degradation across the globe. In China, management of water symbolizes and characterizes the nature of the state responsible for it. How water is treated and how people are treated are bound to be the same, because they draw on the same philosophy. This link between respect for natural waters and respect for the individual has been recognized since antiquity: as the *Guanzi* said in the Spring and Autumn period:

> The solution for the Sage who would transform the world lies in water.
> Therefore when water is uncontaminated, men's hearts are upright.
> When water is pure, the people's hearts are at ease.

If this seems to bode grimly for the political landscape of contemporary China, given how poorly its waters are faring, we can also invert the sentiment to find a more optimistic reading: if the situation of the environment improves, it seems inevitable that so too will the situation of the Chinese people. And indeed, efforts to improve the state of the waters are going hand in hand with a broadening of political discourse and the emergence of new, more liberal voices. The weight of China's impressive store of philosophical and spiritual wisdom, whether it is Confucian, Daoist or the imported but significant tradition of Buddhism, suggests that there should be nothing to fear from this opening up, and everything to gain.

The problems are not insoluble, but they get harder to solve with each passing year. There is no panacea, no quick fix, and solutions will have to be implemented at many levels: scientific and technological, legislative, social and political. Some argue that the answer lies with managing demand, encouraging smarter water use and channelling water of different quality to different applications. Others say that, in China as elsewhere, 'grey' solutions that involve 'hard' human-engineered structures such as dams – which address only some problems, and at the cost of ecosystems, habitats and livelihoods – must be replaced with 'green' infrastructure that uses natural or semi-natural systems such as wetlands, forest ecosystems, rain harvesting and small-scale reservoirs to provide clean water, flood control and water storage for hydropower. But behind these proposals lies the unavoidable fact that to change the waters means to change the country and its people:

there is no answer that is not fundamentally political, whether that means a shift in economic priorities, an admission of new actors to the dialogue, or a devolution of legislative power and responsibility. How China solves its water problem will almost inevitably show us how China is going to evolve more broadly, and perhaps the first will even catalyse the second. And if that problem is *not* solved, the political ramifications will be no less profound.

The war on nature

China lacks a strong tradition of environmental protection, but is in that respect no different to the West. Why, though, did the Confucian recourse to nature as a source of guidance for human conduct not inspire more reverence for nature itself? This is no mystery. That very impulse reflects the anthropocentric bias of Chinese thought: it is all about us, if not indeed all there *for* us. The very landscape is, from peak to valley, engraved with human consciousness: mountains and hillsides not only encode our stories but can be rearranged for our convenience. Active intervention in the environment is indeed one of the aspects in which traditional political philosophy in China differs from that in the West, where the landscape plays no part.

This human-centred view of nature contributes to the appeal of its representation in Chinese art and poetry, as we saw in the previous chapter. The exquisite sensitivity of painters and writers to the contours and energy flows of the natural environment resides in their ability to express how these things evoke mood and sentiment, memory and reverie. The macrocosm is used to explore and depict the microcosm; there is no real tradition of Chinese nature-writing and appreciation in its own right. We can't possibly doubt that the artists and philosophers delighted in nature, but they did so largely because of what it seemed to teach us about how to live.

This isn't to say that an awareness of the value of living in harmony with the environment is absent in Chinese tradition. The Confucian attitude might emphasize management and control, but it also advocated moderation and balance. The *Mencius* recommends practices that will preserve ecosystems: one should not use nets to fish in pools and lakes, and should limit tree-felling for timber to ensure continued abundance. During the Warring States period, a document called the

Wen tao criticized rulers for 'degrading famous mountain spots' and causing great floods. The *Guanzi*, from the same era, warns that 'people who are of ruling quality but are not able to respectfully preserve the forests, rivers and marshes are not appropriate to become rulers'. The other major philosophical traditions were even more inclined to afford the environment respect. Daoists considered that we should adapt to accommodate nature, while Buddhists insisted on a reverence for all life.* 'Humanity follows the Earth,' says the *Dao de jing*, 'the Earth follows Heaven, Heaven follows the Dao, and the Dao follows what is natural.' The environmental campaigner Dai Qing was justified in complaining apropos the Mao era that 'At every turn . . . the Chinese leadership has made decisions which run counter to the Chinese philosophical concepts of maintaining order and balance between humankind and nature.'

Yet the emphasis for China's leaders was generally on taming nature rather than preserving it. There was nothing intrinsically valuable in wilderness: the prevailing view was that it harboured wild, ungovernable peoples. The question was rather whether one approached nature as a gardener, an exploiter, or even an antagonist. After the formation of the People's Republic and the rejection of traditional values and beliefs, nature became regarded not as a wise teacher, nor as a delicate mistress to be revered and 'husbanded', but rather as at best an unruly servant in need of tough discipline and at worst a deadly enemy to be crushed.

Conventional Maoist rhetoric spoke of conquering nature as though it was just another impediment to socialist thought. Quite simply, this was war – and moreover war against a harsh oppressor. 'Man must use natural science to understand, conquer and change nature', Mao said in 1940, 'and thus attain freedom from nature.' The problem was that he was ignorant, perhaps wilfully so, of natural science, which he seemed to regard as an implement that could be shaped by ideology. His former secretary Li Rui, who fell from grace partly through his opposition to the construction of large dams (see page 239), said that Mao was simply irritated when he was told that his plans were

---

* The Alliance of Religions and Conservation, founded in 1995 by Prince Philip of Great Britain, says it is, with the apparent blessing of the Chinese government, 'working with the Chinese Daoist Association, the Chinese Buddhist Association and the International Confucian Ecological Alliance to encourage and support their interest and involvement in protecting the environment'.

incompatible with the laws of nature. He would cite the Foolish Old Man who moved the mountains in defence of the idea that with enough manpower (conscripted or motivated by inspirational propaganda) there was nothing one could not achieve. 'Man's ability to know and change nature is unlimited', he insisted.

The wanton environmental destruction of the Mao era was made all the worse by the urgency with which it was pursued. According to the revolutionary spirit, everything had to be done in a hurry. Even 'rushing ahead' was not enough: the absurd targets had to be met with a great leap forward, a discontinuity achieved with revolutionary fervour. And so entire forests were sacrificed in a single season in order to feed the backyard furnaces that smelted valuable tools and kitchenware into millions of tons of useless scrap metal. These activities were organized along military lines, with peasants formed into brigades, battalions and companies and marching to work singing work songs. 'Everyone is a soldier', said Mao. This mode of work encouraged unquestioning obedience, but the wider agenda was to cultivate a communal sense of struggle and victory: nature became a proxy for the outside forces allegedly threatening China.

All the same, we have seen that there was ample precedent for reshaping nature from imperial times: the Maoist enterprise built on a long-standing culture of environmental manipulation and exploitation. Premier Jiang Zemin acknowledged as much, apparently without reservation, in a speech of 1997 to mark the damming of the Yangtze during the construction of the Three Gorges Dam:

> Since the dawn of history, the Chinese nation has been engaged in the great feat of conquering, developing, and exploiting nature. The legend of the mythical bird Jingwei determined to fill the sea with pebbles, the Foolish Old Man resolved to move the mountains standing in his way, and the tale of the Great Yü who harnessed the Great Floods are just some of the examples of the Chinese people's indomitable spirit in successfully conquering nature.

Given this history, when in 1975 Zhou Enlai announced the Four Modernizations – agriculture, industry, science and technology, and militarization – as an antidote to the disastrous Great Leap Forward, it isn't at all surprising that there was no awareness of how (as the

West had at least begun to discern) modernization poses threats to the natural ecology. In promoting these objectives by opening up to capitalist enterprise, Deng Xiaoping made the priorities plain: 'To get rich is glorious', his slogan read, and the popular reform-era saying *yiqie xiang qian kan* ('look towards money in everything') only followed where Deng pointed.

Guardianship of the environment wasn't entirely neglected, but it took a very low priority. Zhou organized the first National Conference on Environmental Protection in 1973, but it was overwhelmed by the turmoil of the Cultural Revolution. The National Environmental Protection Agency (NEPA) was established in 1984, but its position as a branch of the Ministry of Urban and Rural Construction made its status clear. Even when it was made an independent organization four years later, it was largely toothless. The first Water Pollution Control Law was also introduced in 1984, but even with subsequent amendments it has failed to avert catastrophic industrial accidents or to hold those responsible to account. Economic development should come first, officials insisted – we can worry about the environmental cost later.

Let's be clear: none of this suggests that there is something uniquely bad for the environment about an authoritarian, socialist form of government. That is manifestly untrue, as the rapacious excesses and falsified economic accounting of a capitalist market economy show all too clearly. Shapiro cautions that the link between politics, social justice and the environment applies as much to capitalist countries as it does to China: in the former, nature is seen as a resource to be exploited for profit, and where this happens we are also likely to find people used as a cheap labour force and moulded into a consumerist mass market. Whenever something 'more important' comes before environmental protection and maintenance, whether it is ideology, growth, private ownership or money, then the land, water and ecosystems will suffer. Arguably, the capability for fiat-style governance in China makes environmental problems more feasible to address in principle than is the case in pluralistic democracies, which have shown themselves to be pitifully incapable of taking or even agreeing on effective action against climate change. But it is difficult to see how such changes can ever be instigated without significant intellectual freedom, widespread participation, willingness to adapt and respect for local particularities. And the case of China shows with particular clarity how this change is tied to issues of historical contingency and political symbolism: a question like water management says things beyond itself, it speaks to attitudes of moral and even philosophical relevance.

Running on empty

There have always been times when China's great rivers failed to reach the seas. In AD 309, for example, a terrible drought left the Yangtze, Yellow and Han rivers all dry in their lower reaches. As early as the first century BC there was a debate about how to balance nature's way against human needs. Did excessive irrigation risked violating 'the nature of water', scholars and engineers wondered?

But no one could pretend that, when the Yellow River stopped 650 kilometres short of the Bohai Sea for 226 days during 1997, this was just a recurrence of an old problem. The desiccation was so bad that for a time the Yellow River did not even make it into Shandong province, which relies on the river for over half its irrigation water. The water was simply used up: diverted from the river, or perhaps just never reaching it in the first place. On the North China Plain the agricultural demand for water that would otherwise swell the lower reaches of the river is immense: more than two-fifths of the country's farmland is here, while the annual run-off is just 6% of the nation's total. According to one Chinese hydrologist, the Yellow River has been hit by a double whammy since the 1990s: increasing diversion and consumption of its flow, coupled to a significant reduction in rainfall due to climate change. As a result, the per capita availability of water in the Yellow River basin fell by more than half between 1952 and 2009. Civil engineer Alistair Borthwick, a specialist on China's hydrology, says:

> it appears that the Lower Yellow River's ability to meet its ecosystem and socio-economic requirements is exhausted . . . the Lower Yellow River is not sustainable at the level it had in the 1950s, and has declined to the point that it is unable to meet any of its primary functional requirements even after the implementation of recent countermeasures.

Ever since China's economy began to grow in the late 1970s thanks to the reform policies of Deng Xiaoping, water use had risen in concert.

In part this was driven by the rise in manufacturing, for industry is a thirsty business. But agriculture accounts for almost two-thirds of water use in the country: 70% of China's food relies on irrigation. And whereas a resource like water often suffers from a tragedy of the commons across the globe, with users figuring that they might as well overexploit it like everyone else, in China it has become ever less clear that there is a commons in the first place. As land began to fall into quasi-private ownership, there was little sense of the communal obligations that, for all their distortion and abuse, had at least been observed to some degree during the Communist era.

While regulating water use by pricing mechanisms brings its own complications, state guarantees of water supply to rural areas has reduced any incentive to improve the efficiency of irrigation, and the rise in living standards has led to increased consumption of meat and the expansion of water-inefficient animal husbandry. Now that farmers are free to sell surplus product on the market, they are eager to increase productivity by using more water, and they prefer to grow high-value, water-intensive crops such as fruit and nuts rather than grain and potatoes. The demand for water from industry and urbanization in north China has risen too: the proportion of water resources used by industry has more than doubled since the 1980s.

On the North China Plain and other northern and western regions, exhaustion of surface water means that irrigation relies increasingly on tapping groundwater via wells. As a result, the water table is falling in some areas at an alarming rate – up to a metre a year. This in turn triggers subsidence. Roads have disappeared into sinkholes in Beijing and there are concerns that the international airport could be threatened. Even Shanghai, situated in a relatively moist climate zone, has sunk by around two metres in the past fifteen years as underground water is depleted.

When land grows arid, it may go into irreversible decline: farmland turns to desert. In a symbolic blow to China's national prestige, the Great Wall itself is being overcome by sand. Excessive use of water for irrigation in the north-west province of Gansu has led to a drying up of oases in the Shiyang River valley in Minqin county, leaving a section of the Great Wall there, dating from the Han dynasty, exposed to creeping desertification and the erosive effects of sandstorms. Some parts are already buried, others – made from mud rather than brick and stone – have cracked and crumbled. Meanwhile in the provincial capital of Lanzhou sandstorms leave the sky eclipse-dark, and everything takes on the dirty tan of a sepia photograph. When these storms hit Beijing they can virtually paralyse the city, and there was an ominous resonance with past invasions from the north in Premier Zhu Rongji's suggestion in 2000 that the capital might need to be moved as a result. Meanwhile, loss of farmland has forced massive migration and relocation of peasant populations: around 20–30 million farmers in the north-west province of Ningxia have been displaced following desertification and degradation of the land, and hundreds of millions more have been adversely affected by it.

Problems of desertification, erosion and salinization due to over-cultivation and excessive irrigation have been made worse by China's urban growth, and not simply because cities need feeding. In 1994 the government decreed that all cultivable land taken over for construction had to be offset by reclamation of cropland elsewhere, lest the country lose its food self-sufficiency. As with so many market-based 'balancing' mechanisms, this just shifted the problem elsewhere. Rich provinces paid poor provinces to develop grasslands, leaving the newly cultivated soils exposed to run-off and erosion.

Taking water north

The north–south axis that oriented the Grand Canal continues to dominate Chinese water conservancy today. But now it is not tribute grain that the northerners need, but water itself. The extensive and sometimes unnecessary irrigation schemes of the Mao era (see page 225) drained the Yellow River and its tributaries, while dams retain much of the flow in massive reservoirs. It's no wonder that the river began to run dry.

To alleviate such problems, the idea arose during the technocratic 1950s that Yangtze water might be brought north along channels comparable in scale and ambition to the Grand Canal. Mao suggested that, 'if at all possible', it would be good for the water-strapped north to 'borrow' from the south. Four possible routes were considered for the *Nan shui bei diao* ('Southern Waters Transported North') scheme, from deep in the mountains of Sichuan to the coastal east from Shanghai through Shandong.

But nothing concrete came of these plans until 2004, when the government authorized the South–North Water Transfer Project. It aims to replumb China by transporting 45 billion cubic metres of fresh water every year from the humid Yangtze basin along three separate canals each over 1,000 kilometres long to the North China Plain.

The eastern route follows the Grand Canal along much of its course. The existing channel here has been deepened and broadened, and includes dozens of pumping stations along the way to raise the water up the incline from the Yangtze to the Yellow River. The central route draws water from the immense Danjiangkou Reservoir in Hubei, formed by a dam on the Han River that was constructed during the Mao era but has been raised as part of the water-diversion scheme. This channel carries water all the way to Beijing, where it is forecast to satisfy a third of the annual water demand. In an extraordinary subversion of geography, both the eastern and central routes require the flows to be directed along tunnels *under* the Yellow River. The western route is technically the most demanding, however, since it has to negotiate mountain ranges on the Tibetan plateau, via tunnels and dams, to carry water from the Yangtze headwaters to those of the Yellow River. Some doubt that it will ever be built. The whole scheme has so far cost around $79 billion, which is more than twice the sum spent on the Three Gorges Dam.

The eastern and central routes are now operating. Although hopes that the scheme would deliver water to Beijing in time for the 2008 Olympics were dashed, it began to supply the capital with southern water in 2015. Like all large-scale hydroengineering projects, this one has been divisive. Some Chinese citizens are unhappy that they derive no benefit from a flow that crosses their land behind high concrete walls. Others have been displaced to make way for the channels and for the raised waters of Danjiangkou Reservoir. And local officials in Hubei province say that the sapping of resources from this storage lake and from the Han River will compound their own problems of water supply and quality.

Water resources have also been lost by reclamation of land for farming through construction of polders and draining of lakes. This

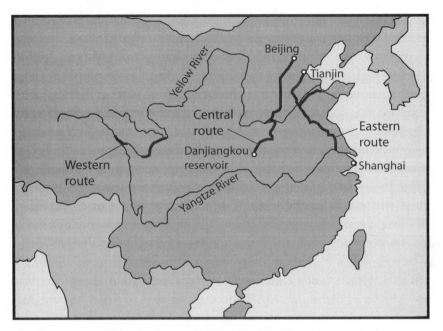

The three routes of the South–North Water Transfer Project.

was happening at least as early as the Song dynasty, when the official Wei Jin complained:

> The people were deprived more and more of the benefits of reservoirs and lakes. What used to be rivers, lakes and marshes has all become land in the last thirty years . . . The superiors (officials) and inferiors (powerful families) shut their eyes to each other's acts and pretend not to know that anything has happened.

Not only has this practice been pursued aggressively and sometimes without any state oversight, but it has also paid scant regard to the realities of cultivation, so that some reclamation projects produce only useless, barren earth. Unrealistic targets for agricultural production during the Mao era led to the draining of lakes motivated by the slogan *wei hu zao tian*: 'encircle the lakes [to] create fields'. In once lake-rich Hubei, half of the main water bodies were filled and the total lake area was reduced by three-quarters. One of the most disastrous projects was the draining and filling of the great Dianchi Lake and the surrounding wetlands near Kunming in Yunnan, which wreaked havoc with the aquatic and marshland ecosystems, left the remaining lake waters filthy, and seems potentially to have altered the local microclimate. Once famed for the mildness of its 'four seasons like spring', the Dianchi region now seems to be experiencing harsher summers and winters, thought to be due to reduced evaporation from the wetlands.

In this misguided project, the connection between Mao's vision of remoulding nature and remoulding people – of the sociopolitical implications of mastery over water – was made explicit. As the draining and poldering continued apace, in 1970 the newspaper *Yunnan ribao* quoted the Chairman as saying, 'To reform the objective world is also to reform your own subjective world', and went on to explain that 'To struggle with the heavens, struggle with the earth, struggle with the waters, struggle with class enemies, and struggle with all kinds of wrong ideas is powerfully to promote a revolution of thought.' Thus, building polder dams at Dianchi was to build a wall 'against revisionism in our minds'; to drain the lake was to expel 'the muddy waste-water of capitalism from the deep part of our souls'.

Lakes have always been revered in China as places of beauty, repose and meditation. Now, if they are not mired in green slime and do not stink of dead fish, they might not be visible at all. Lake Poyang on the Yangtze in Jiangxi has been in retreat over the past decade or so, the dry season commencing about five weeks earlier than it did on average in the second half of the twentieth century. Fishing boats are left stranded on cracked mud, and the fishermen are forced to seek a livelihood elsewhere. In 2007 the lake fell to the lowest level ever recorded, leaving hundreds of thousands of people without adequate water. Many suspect that these changes in the great Yangtze lakes – Dongting has contracted too – are linked to the restriction of flow by the Three Gorges Dam. Yet while Poyang has indeed grown smaller since the Three Gorges Dam was completed, it was already shrinking anyway because of excessive water withdrawals and because the climate had become drier. Lake Baiyangdian in Hebei province, the largest in northern China, is also shrinking, some say because of the damming projects of the 1950s (although land reclamation here for farming began during the Qing era).

As China's water resources dwindle, climate change is the joker in the pack. The issue is not whether it will make things worse, but by how much. By the latest reckoning, the glacier volume on the

Qinghai–Tibet plateau, which provides the headwaters not just of the Yellow and Yangtze rivers but also of the Mekong, Salween and other major rivers of South-East Asia, is predicted to shrink by up to 27% by 2050. In Ningxia, Xinjiang and Qinghai, a reduction in run-off of 20–40% per capita is predicted by the end of the century. China's water resources already fall short of its needs by 40 billion tonnes a year, and this is predicted to rise to 58 billion by 2020. The Ministry of Water Resources predicts that there will be a 'serious water crisis' by 2030. And if China runs out of water, its economy will stall.

The solution to water depletion is far from clear. In many parts of the world such problems have been addressed by pricing mechanisms: if you want people to use less water, charge more for it. There is certainly some ground for supposing that while farmers pay next to nothing for irrigation water they have little incentive to invest in more efficient technologies for delivering it. And industries can be similarly wasteful and polluting instead of developing strategies for water recycling. But water use is one of those areas in which a trust in market mechanisms to drive efficiencies is naive. Allocating water only to those who can afford it is a recipe for severe social unrest.

Many Chinese people seem to have a sense both of liberating optimism and of immense hazard in mega-engineering techno-fixes like the South–North Water Transfer Project. Only the state has the means to offer them security in the face of nature's terrible threats, but there is a price to be paid – in rice, in labour, in freedom.

It will be far better, say some critics, to address the root of the problem by managing demand, especially by introducing more effective means of conserving and recycling water, reducing pollution, and matching quality to application. Relying on centralized treatment plants to restore all water to drinking quality imposes an unnecessary burden: most water used in municipalities is for industry, agriculture and construction, and even in households it does not all need to be potable. Small-scale purification at the point of use might therefore make more sense – but not with the unregulated and unreliable devices currently on the Chinese market. These options and others are already widely discussed in the light of what threatens to be a global water crisis around the middle of the century. But they are of particular urgency in China, where precipitous economic growth coupled to specific geographical and historical contingencies of water management have created a more or less unique situation.

Dirty water

On 13 November 2005, 10,000 people in Jilin City, Jilin province, were evacuated from their homes after an explosion at the No. 101 Petrochemical Plant on the bank of the Songhua River. Six people were killed in the blast and scores were injured. The evacuations, enforced while the harshness of the northern Chinese winter was approaching, were prompted less by the risk of further explosions than by fear that the population would be exposed to the toxic and carcinogenic products of the plant. As it was, around a hundred tonnes of benzene, aniline and nitrobenzene found its way into the Songhua, a tributary of the Amur – the river also known as Heilongjiang ('Black Dragon River'), which flows through the neighbouring province and gives it its name. The slick, smelling of sweet almonds and rotting fish, was eighty kilometres long by the time the Amur reached Heilongjiang's provincial capital of Harbin. The water supply for the city's 4 million inhabitants was temporarily suspended, and the clean-up process cost the equivalent of around $2 billion.

China has become resigned to these 'black dragons' in its waterways. Just three months after the Songhua incident, 20,000 people in Yibin prefecture at the junction of the Min and the Yangtze went without

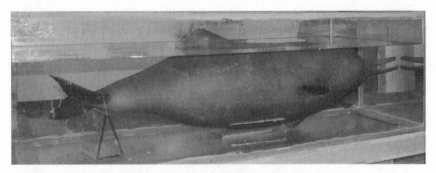

The Yangtze dolphin or *baiji* is thought now to be extinct. This specimen is preserved at the Institute of Hydrobiology in Wuhan; a line of scars left by a rolling hook is visible on the centre of the body. A sculpture in Tongling, a Yangtze river port in Anhui, (*opposite*) testifies poignantly to the affection with which these creatures were once regarded.

water when a spill from a chemical plant polluted the Yuexi River, a tributary of the Min, with fluorine, phenol and other toxic chemicals. Mining and metalworking are among the worst culprits, as they always have been. In Duqibao village in Yunnan, residents were left dizzy and vomiting in 2008 when their water supply was contaminated by a mining company. That same year, people in Jianli county, Hubei, suffered skin rashes from the effluent of vanadium smelting, and in Guangxi province arsenic flushed by rain from the waste dumped by a metallurgical company poisoned a local village.

The decline of the waterways is in no respect more poignantly symbolized than by the probable extinction of the Yangtze dolphin, the *baiji*, said to be the descendant of a princess who fell in love with the river god. These elegant creatures were deified for centuries as a kind of mermaid. But dams have obstructed their access to upstream breeding grounds, overfishing has robbed them of their food, the raised silt loading of the waters through deforestation makes it hard for them to see, and the pollution poisons them. Worst of all, Mao's rejection of old beliefs during the Great Leap Forward – 'There is no Heavenly Emperor or Dragon King', he proclaimed – removed the taboo on hunting the *baiji* itself. They were caught for their meat and oil, and their hides were used to make bags and gloves. Ecologists now believe that the *baiji* are already gone forever.

Baiyangdian, the shrinking great lake in Hebei, was once known as the Bright Pearl of northern China, a place of reeds and mist populated by mandarin ducks and wild geese. Its rich flora and fauna have provided a living for the local people, and its role in sustaining ecosystems and providing flood storage has won it the more prosaic title of the Kidney of North China. But thanks to an intake of toxins the region is now facing the prospect of kidney failure. Between 1988 and 1992, chemical and petrochemical plants and paper mills in the city of Baoding close to the lake shore discharged effluent into the waters, mixing with raw sewage from the city. Fish began to die off, and fishermen lost their livelihood. The lake water is now undrinkable.

There are problems everywhere you look. Lake Tai (Taihu) in Jiangsu province, China's third-largest freshwater lake, suffers annually from blooms of blue-green cyanobacteria: algae that excrete toxins harmful to the liver, gut and nervous system, and which are nourished by nitrogen and phosphorus in the discharge streams of chemical and manufacturing plants, or in fertilizers leached from the soil. The problem reached crisis point in 2007, when a particularly big bloom overwhelmed the treatment plants supplying clean water to the city of Wuxi on the lake's northern shore. Thanks to rotting masses of

What hope is there for China's once 'sweet' river waters, lined now with factories discharging their effluent with little effective regulation?

algae, the water came out of the taps smelling like putrid fish, and wasn't safe to drink even if you could ignore the disgusting smell. Some two million people were left without drinking water on tap for a week.

Water-management practices such as damming may make these problems worse, as dams and locks can become holding basins for contamination. In July 2001, heavy rains in the Huai valley flushed out 173 billion litres of polluted water held behind lock gates on the upper river, turning the waters foul downstream. There are around 4,000 such traps along the river, and they inhibit the natural flushing of pollution. The Huai valley, like other catchments, suffers from a lack of coordination among regional authorities – the river flows through four provinces – and from weak environmental oversight. An episode of public poisoning in 1994 following the dumping of toxic waste into the river from factories led to fresh efforts to hold the industrial polluters to account, but attempts to control them were easily evaded. Some factories simply started operating under cover of night, others received tip-offs about inspections and so were able to shut off waste outlets in advance. Local authorities generally favoured economic development over environmental curbs. As a result, local residents complained that the Huai water was unfit even for washing clothes, and that it put pigs off their food. It's the same story more recently: in 2008, over 60% of the Huai water was found to be unsuitable for fishing or drinking, and in some cases too polluted even for agricultural use.

All of China's watersheds, and almost half of its water sources, are now badly polluted by fertilizers, pesticides and heavy metals from mining and industry. More than 300 million of its citizens live without access to safe drinking water. Even after treatment, 40% of the nation's water is fit only for industrial and agricultural use; a quarter of waste-treatment plants fail the quality controls. Regulation of chemical plants is very weak, and government agencies that should nominally protect water quality in reality have very little power. Many of the waste-water treatment plants are ineffective; some stand idle. Thanks to contaminants such as arsenic, fluoride and toxins from untreated waste-water and leaching from land-fills, a third of Chinese cities failed to meet drinking-water quality stand-ards in 2006. Cancer rates are reported to be rising, and one leading researcher at the Chinese Academy of Science's cancer research institute says that this is because 'pollution of the environment, water and air is getting worse by the day'. Every year, it is estimated that water pollution in China produces 190 million casualties and around 60,000 fatalities.

The notoriously poor air in major urban centres poses a hazard to human health and can make life miserable. Yet life can and does go on in this noxious atmosphere. But without adequate water, China's burgeoning economy could come to a standstill. Already, pollution and destruction of the environment are estimated to be costing China around a tenth of its gross domestic product. In 2007 Zhou Shengxian, head of the State Environmental Protection Administration (SEPA, now the Ministry of Environmental Protection), stated that 'Serious water pollution has affected people's health and social stability and become the bottleneck thwarting China's sound and rapid economic and social development.' Even President Hu Jintao – like several of his predecessors a hydraulic engineer by training – commented in late 2012 that China's development is 'unbalanced, uncoordinated and unsustainable'. That warning has become almost a reflex among China's leaders, but it is not simply a pretext for instituting more government control. There is real concern at the parlous state of the country's water.

Wave of anger

The Chinese government fears not just that water problems will slow down economic growth and the rise in living standards that citizens have come to expect during the past few decades, but that they will destabilize the whole society. It's a justified fear. Disputes over water use and management have already flared into violent conflict. Water rights on the Zhang River, a tributary of the Hai that flows through Shanxi, Henan and Hebei, have sparked pitched battles between villagers, including fatal mortar attacks and bombings.

Meanwhile, a protest in 2004 against the hydroelectric Pubugou Dam on the Dadu River in Sichuan led to a riot quelled by thousands of police, during which one of them was killed. For this crime one of the protestors, Chen Tao, was subsequently executed, and another was jailed for life. The stakes were raised by the historical and symbolic significance of that site: the dam is close to Luding, where in 1935 the Red Army crossed the river in a conflict with the Kuomintang forces that might otherwise have seen it defeated, and the course of Chinese history thereby altered.

The Pubugou demonstration was the biggest civil protest movement since Tiananmen in 1989. It was a chaotic affair with many disparate voices, but nevertheless expressed such deep dissatisfaction that both Hu Jintao and Premier Wen Jiabao were forced to concede that the protests over relocation of peasants should be heeded. Nevertheless, construction of the almost 180-metre dam went ahead, farmland was submerged, and more than 100,000 people were displaced.

It has not always turned out that way. Damming of the Min River in Sichuan was on the cards since the Sichuan party secretary Li Jingquan was embarrassed by Mao's inability to swim the river during a visit in the 1950s, because of its turbulent flow. Construction of the hydropower Zipingpu Dam on the Min finally began in 2001 and was completed in 2006.* Yet when it emerged in early 2003 that a second dam was being planned nearby at Yangliuhu, protestors were able to mobilize opposition by arguing that the nearby ancient waterworks of Dujiangyan, masterminded by the water hero Li Bing, would suffer. In August 2003 the plans were abandoned – an unprecedented victory for opposition to a scheme given government approval.

Plans for a series of hydroelectric dams on the Nu River in Yunnan were also shelved by Wen Jiabao in 2004, following a canny campaign by the environmental groups Green Watershed and Green Earth Volunteers. The protest focused on the relocation of ethnic minorities whose ancestral homes were in the region, and on the potential harm that could be done to the rich and diverse ecosystem in what is now a UNESCO World Heritage site. Dams, objectors claimed, might disrupt the spawning grounds of endangered fish species. The project was fraught internationally too, for the Nu is the Chinese name for the Salween, which flows on through Myanmar and Thailand and provides water and fish for millions of peasants in those countries.

While this decision was hailed as a victory for China's environmentalists, it has proved temporary. Shortly before he retired in 2013 – some regard the timing as no coincidence – Wen reinstated plans for five of the thirteen dams originally planned for this stretch of the Nu. Now construction has begun, despite claims that there is insufficient understanding both of the geological and engineering challenges and of the possible

---

* The Zipingpu Dam has not lacked controversy, being possibly implicated in the devastating 2008 earthquake in Wenchuan (see page 245).

environmental consequences in an area that hosts endangered species such as the red panda and the snow leopard. The homes of ethnic minorities living on the steep, forested slopes of the Nu valley are threatened too.

A similar see-sawing uncertainty hovers over the famous Tiger Leaping Gorge on the Jinsha River in Yunnan, where these headwaters of the Yangtze crash and tumble through a precipitous canyon. Travellers and backpackers have made the arduous journey here for decades to witness the spectacular scenery and to enjoy the hospitality of the Naxi people who live in the region. In 2004 the Chinese government proposed to dam the gorge for hydroelectric power, submerging the slopes beneath a placid reservoir and displacing the Naxi, perhaps northwards into the harsher heights of the Tibetan border. These plans were abandoned by the Yunnan provincial government at the end of 2007 – but that has not prevented the commencement of construction for several other dams in the region, just outside of the area designated as a UNESCO World Heritage site. Whether the legendary leap of a tiger across the narrowest point of the gorge (twenty-five metres) to escape a hunter will continue to be feasible in the future remains unclear.

Towards pluralism?

If China is going to be able to deal with the environmental problems that have followed in the wake of its extraordinary economic growth, the impetus will come less from a sense of obligation to nature than from an awareness of the human cost: of soaring cancer rates, of rivers filthy and foaming with poison, of declining agricultural yields and of protests from populations forced to live in unliveable places. With twenty of the thirty most polluted cities in the world lying within the Chinese border, three-quarters of the country's water undrinkable and a quarter of its cultivable land afflicted by desertification, salinization or serious soil erosion, it is obvious even to the most technophilic leader that things can't go on this way. 'The environment will be the area in which many of the crucial battles for China's future will be waged', says political scientist Elizabeth Economy.

Already, protests and controversies over water are playing a central role in changing the political climate of the country. There is nothing the Chinese government fears more than domestic unrest – and it seems that there is nothing rural communities will oppose more fearlessly than threats to their access to land and water. So deeply are issues of water access felt in China that they have the potential, in the words of Canadian political analyst, Nathan Nankivell, to 'present a unifying focal point for dissent that crosses geographic, cultural, socio-economic and political lines'. The head of the SEPA, Zhou Shengxian, reported that in 2005 there had been an average of 1,000 environmental protests in China every *week* – a level of dissent that no purely political causes can match. 'Pollution', said Zhou, 'has become the "primer" for social instability.'

Some of the first truly effective non-governmental organizations* in China have been motivated by water and related ecological issues. Despite the occasionally harsh official response to protests, some of these organizations have not only been more or less tolerated but even heeded, as in the case of the Yangliuhu Dam in Sichuan. NGOs such as Friends of Nature, formed in 1994 by activist Liang Congjie, and Green Watershed, founded by environmental scientist Yu Xiaogang and focusing on water-management issues, have found a way to challenge the party position on the environment without provoking an authoritarian response. Another prominent NGO is the Institute of Public and Environmental Affairs (IPE), set up by writer Ma Jun to 'promote effective participation in governance'. Ma is the author of *China's Water Crisis*, an exposé of river pollution that has been compared to Rachel Carson's *Silent Spring* (though if it has comparable poetic qualities, they are lost in translation). The IPE operates a database on pollution, and seeks to mobilize consumer choice to ensure that industries observe their environmental obligations.

According to David Pietz, a leading historian of China's environmental affairs, these organizations are keenly aware that their effectiveness and indeed their very existence is contingent on recognizing the limits of their powers. They operate not so much within the political system as within its fissures – which is enough to give them some degree of political leverage. It is only in this liminal space, say some environmental specialists, that there is room for an open-minded exploration of the scientific issues. Yu Xiaogang quit his position in a government ministry in Yunnan to found Green Watershed in Kunming,

---

* Chinese NGOs are not exactly that – to be granted official legitimacy they must have a sponsoring governmental department. This does not, however, appear to be greatly compromising their autonomy, even if it means they must tread carefully.

which led the opposition to damming of the Nu River. 'Inside the system, you can do only so-called "decision-making supporting research'," he has said. 'That means the government has already made the decision. You do research to support the decision. You never do something that changes the decision.' Yet Yu stresses that he and his organization are aiming not to 'protest' or to oppose the government, but to 'construct a dialogue' – through which, for example, hydroelectric companies will acknowledge their risks and obligations.

The government knows that it is in its interest to permit and even encourage environmental advocacy. It knows all too well that unofficial 'green' pressure groups acted as lightning rods for social rebellion and demands for reform that led to the collapse of the Communist authorities in the former Soviet Union and the Eastern bloc. Officially sanctioned environmental groups in China don't just provide a pressure-release valve, however. They also fill a gap in environmental enforcement, relieving the central government from the onerous task of trying to hold local polluters to account. The government's top-down approach to environmental management – grand campaigns to stimulate green initiatives or raise awareness, coupled to a reliance on market forces and private enterprise to actually put ideas into practice – carries rather little weight at the local level. The laws to prevent pollution and promote green policies tend to be drawn broadly but sketchily, and are posed more as desired consequences without a clear legal framework for achieving them. For individuals or communities seeking to use such laws to protect their environment and livelihoods, it is not at all clear what mechanisms they should follow. Local courts can in any case interpret the laws in different ways. This, indeed, is the paradox of China, where a 'central planned economy' doesn't at all equate with intrusive government, but quite the opposite: a great deal is left to the discretion of local authorities, for better or worse. The party doesn't *want* this kind of power, any more than it wants the burden of social welfare for the entire population.

And so there is little at the legislative level to stop further degradation of China's land, air and water. Fines for pollution are scarcely a deterrent: it is often economically preferable for companies to pay up rather than clean up. The penalties don't cover the true costs of clean-up, and they are usually one-off payments even if the infringement is repeated. Environmental stewardship by the government remains patchy. When NEPA became SEPA in 1998, its staff was virtually halved in the process. In 2008 SEPA became the Ministry for Environmental Protection, for the first time making it a cabinet-level administration with a vote on the State Council. Yet by all accounts its staff are still under-resourced and overworked.

Even a new environmental law introduced at the beginning of 2015, said to be the 'most progressive and stringent in the history of environmental protection in China', has been criticized as too lenient. It introduces penalties for manipulating environmental data and makes local agencies and authorities more accountable, but suffers from the old problem that its safeguards can be challenged and overruled by other departments with conflicting priorities. Political scientists Zhang Bo and Cao Cong complain that the new law 'fails to acknowledge citizens' basic right to an environment fit for life'. Some argue that such legislation needs to be afforded the highest level of priority, on a par with laws directed at promoting economic growth and controlling population.

The widespread claim that China's water will decide its future therefore amounts to much more than the truism that an industrial and social economy relies on its water resources. Water seems likely to reshape the entire political landscape – and in a way that does not conform to any Western model of a post-authoritarian state. It seems possible, at least, that the hundreds of thousands of Chinese NGOs now officially registered, for all their limited scope and freedom, might be the seeds of broader change in the relationship between the Chinese state and society. In one view, grass-roots water movements offer a glimpse of what the future political face of China will look like: not exactly democratic, but pluralist, where a 'fragmented authoritarianism' permits high-level decisions to be malleable in the face of public opinion as they percolate down the chain of command.

Some activists already sail close to the wind in their desire to see something resembling democracy emerge, in whatever form that might take – not so much because this is a fundamental political right, but because it is the only way to achieve true harmony with the environment. Journalist Tang Xiyang, whose wife died of injuries from beatings by Red Guards during the Cultural Revolution, has written that 'I found the chief guarantee of nature protection to be the practice of democracy. Without real democracy there can be no everlasting green hills and clear waters.' The philosopher He Bochuan, an opponent of the Three Gorges Dam in the 1980s, agrees that only political reform can save the environment. Whether or not that is so, it seems incontestable that you can neither run a prosperous and efficient market economy nor protect the environment without qualities such as transparency, respect for (and ability to implement) the law, and openness to domestic and international expert opinion.

The fact that water resources and management in China are becoming a bellwether of political sentiment is precisely what the history of water in China should lead us to expect. The ancient, shared roots of both water philosophy and water management in China mean that this is an issue on which the country's leaders ignore the wishes of the people at their peril.

Acknowledgements

Why we're drawn to particular places is often a mystery. To the Westerner, China can seem forbidding and harsh, remote, baffling and frustrating. But I was still in my teens when I fell under its spell, and when the opportunity later arose for me to spend some time travelling, China was my immediate choice of destination. I don't know if I found what I expected, based as it necessarily was on romantic imaginings. But circumstances have conspired to take me back to China many times since, and I shall continue to return. It has been my immense privilege to meet many Chinese people who have deepened my appreciation for their country; some are now good friends and colleagues. It is first to all of these folk that I offer my gratitude and acknowledgements: to their spirit, their warmth and generosity, by which I am humbled.

It was a leap of wild optimism to attempt a book of this sort, and the faith and trust offered by my publishers has been essential. For this, and for their diligence in seeing the project through, I am deeply grateful to Stuart Williams and Jörg Hensgen at the Bodley Head, and to Karen Merikangas Darling at the University of Chicago Press. My agent Clare Alexander was equally brave to represent it, and as ever her advice has helped me to find some kind of shape and structure for what is really a rather absurdly overwhelming subject. Several people kindly provided advice, publications and images – in particular, Sarah Allan, David Clarke, Jan Engberts, Li Cho-ying, Sam Turvey and Thomas Stephens.

My family has shared in more of this process than I have usually asked of them, braving the Yangtze drizzle and Henan heat with humour and resilience, and with not much more than dumplings by way of recompense. I hope it was worth it, for I am quite certain that more adventures await us in the Middle Kingdom.

Philip Ball
London, June 2016

Picture Credits

Notes

Introduction

'I repeat that everything appertaining to': *The Book of Ser Marco Polo the Venetian concerning the Kingdoms and Marvels of the East,* trans. and ed. H. Yule, 3rd edn, revised by H. Cordier, 1903, Vol. II, John Murray, London, p. 190.

1 The Great Rivers

Epigraphs: 'The wide, wide Yangtze': Lynn, p. 22; 'The sun goes down': translator unknown.

'twisting around ten thousand times': Allan (1997), p. 24.

'A great man should in the morning': Ward, p. 204.

'He would travel with a servant': ibid., p. 47.

'read more like those of a twentieth-century': Needham & Wang, p. 524.

'The rise of the wave follows': ibid., p. 488.

'Clouds and rain are really': ibid., p. 468.

'Now the rivers in the earth': ibid., p. 487.

'the worst mistake in the history': J. Diamond, *Discover*, May 1987, p. 64.

'an infinite variety of queer': Clark, p. 39.

'Inconceivably great are the benefits': Needham et al. (1971), p. 378.

'divine perspicacity': ibid., p. 280.

'A China without such an immense': Winchester, p. 3.

'All the solitary hills': Watson, pp. 102–3.

'crowds of young lads': ibid., p. 87.

'As we tacked along the Great River': ibid., p. 123.

'The rich wash of scarcely explored': Lynn, p. 3.

'carrying with it numbers of junks': ibid., p. 84.

'I stood on Purple Mountain': ibid., p. 42.

'Witch mountain is high': Schafer, p. 84.

'some vying one with another': Watson, p. 176.

'We were then on what looked': ibid., p. 177.

'Tracking a 120-ton Mayangtzu': Van Slyke, p. 121.

'A big junk of 150 tons': Watson, pp. 161–2.

'the Yang-tse is not only the main': Chetham, p. 88.

'On every man almost': Van Slyke, p. 125.

'The *laopan*, or skipper': ibid., p. 123.

'I think even the most restless person': Lynn, p. 18.

'a boundless expanse': Watson, p. 104.

'celestial looking glass': ibid.

'Vast, vast: lose all sense of direction': Murck, p. 104.

'From the tower': ibid., p. 105.

'Ah, river – I love your robust strength': Lynn, pp. 265–6.

'A bridge will fly to span': Shapiro (2012), p. 95.

'To draw an analogy': *The Writings of Mao Zedong 1949–1976: January 1956–December 1957*, eds M. Y. M. Kau & J. K. Leung, M. E. Sharpe, 1993, Routledge, London, p. 630.

'It is big, but not frightening': *Poems of Mao Tse-tung*, ed. H.-L. Nieh Engle & P. Engle, 1974. Dell, New York, p. 94.

'This is the greatest happiness': see https://iconicphotos.wordpress.com/2009/04/25/mao-swims-in-the-yangze/

'If you would just work at it an hour': *The Writings of Mao Zedong, 1949–1976: January 1956–December 1957*, ed. J. K. Leung & M. Y. M. Kau, 1992. M. E. Sharpe, Armonk, NY, p. 630.

2 Out of the Water

Epigraph: Lewis, p. 53.

'We passed Lion Promontory': Watson, p. 100.

'We then went to the Monastery': ibid., p. 113.

'that seize attention': Blumenberg, p. 34.

'abruptly told without any sense': Wu, pp. 1–2.

'Beware the songs of Chu': Van Slyke, p. 46.

'The people are lazy and poor': ibid.

'Within China a small group': C. Coonan, 'A meeting of civilizations', *Independent*, 27 August 2007. http://www.independent.co.uk/news/world/asia/a-meeting-of-civilisations-the-mystery-of-chinas-celtic-mummies-5330366.html

'chanced to obtain the Way': Dundes, p. 42.

'the people rejoiced': ibid., p. 86.

'a charter for social action': ibid., p. 3.

'Probably no other people': Needham et al. (1971), p. 247.

'every aspect of the Chinese': Lewis, p. 20.
'Like endless boiling water': Wu, p. 69.
'Kung Kung abandoned the Way': Dundes, p. 98.
'When he heard a single good word': Lewis, p. 34.
'If it hadn't been for Yü': see Clark, p. 40.
'violent winds': Wu, p. 76.
'Shun gave the Chinese a heritage': ibid., p. 81.
'The flood myths that developed': Lewis, p. 152.
'abusive and disputatious': Wu, p. 70.
'This is not permissible': Lewis, p. 40.
'Thinking retrospectively': Li, p. 61.
'the sanctioning power of myths': Pietz (2014), p. 31.

3 Finding the Way

Epigraph: Needham et al. (1971), p. iii.
'from an ample source': Allan (1997), p. 3.
'As the years went by': Needham et al. (1971), p. 268.
'The intelligent find joy in water': Allan (1997), p. 23.
'we could think of it': ibid., p. 88.
'Some Chinese terms': Introduction to Chan, p. xi.
'Water is the blood': Allan (1997), p. 123.
'The *dao* is like a vessel': Creel, p. 113.
'Water, which extends everywhere': Allan (1997), p. 24.
'Fish go to one another in water': ibid., pp. 78–9.
'I hear that in Chu': Graham, p. 174.
'Do the heavens revolve?': Creel, pp. 111–2.
'Man is water': Watts, pp. 48–9.
'It is accumulated in Heaven': ibid., p. 48.
'It is only he who knows': ibid.
'The sage's transformation': Priscoli & Wolf, p. 1.
'Seldom his posterity': Keay, p. 69.
'In funerals and ceremonies': Creel, p. 45.
'Perfect balance is found': Palmer, p. 43.
'Heaven sends down': Creel, p. 73.
'man's nature is like whirling water': Chan, p. 52.
'Water, indeed, is indifferent': ibid.
'because he had a better understanding': Allan (1997), p. 42.
'The nature of man is evil': Creel, p. 132.
'the voice of a jackal': Keay, p. 79.
'Each of the Five Elements': Needham & Ronan, p. 144.

'Mercury was used': Keay, p. 104.

'The reason that the River': Allan (1997), p. 45.

'all of the disgusting things': ibid, p. 47.

'If water is still': ibid., p. 53.

'There is no one in the world': ibid., p. 48.

'if the Way is cultivated': Chan, p. 116.

'Where there is a channel': Allan (1997), p. 24.

'One who is good at overcoming': ibid., p. 139.

4 Channels of Power

Epigraph: Needham et al. (1971), p. 378.

'Even if China were not of itself': Le Comte, pp. 101–2 (modernized translation).

'If this is true': ibid., p. 103.

'a Water Conservancy Office': Needham et al. (1971), p. 223.

'chose us and gave us': Creel, p. 31.

'The Mandate of Heaven is not easily': Chan, p. 7.

'The water question conditioned': Chi, p. 73.

'it was the task imposed': Wittfogel, p. 13.

'after having destroyed the nations': Boulanger, p. 23.

'interference of a centralizing': K. Marx, 1853. 'The British. Rule in India'. *New York Daily Tribune*, 25 June, p. 5.

'the development of public works': Chi, subtitle.

'has long been discredited': Burke & Pomeranz, p. 121.

'the general contours': Mertha, p. 1.

'Agriculture is the basis': Needham et al. (1971), p. 264.

'Wherever the canals passed': Chi, p. 66.

'Above, high beacons of rock': Van Slyke, p. 56.

'Upon the side of some mountains': ibid.

'Go by boat in the south': Elvin, p. 136.

'a main artery to bring tax grain': Needham et al. (1971), p. 227.

'By dredging the Qi River': Xiong, p. 93.

'These water-passages, as they call them': Le Comte (1739), p.106.

'wing and waves are her village': Elvin, p. 136.

'The Great Khan has made': Needham et al. (1971), p. 312.

'strong and wide embankment roads': Van Slyke, p. 73.

'The city is all bridges': Keay, p. 412.

'it rules Chinese taste': L. Cooke Johnson (ed.), 1993. *Cities of Jiangnan in Late Imperial China*. SUNY Press, Buffalo, p. 84.

'it is bad form to show any interest': Lynn, p. 7.

'unbelievably *alive* with all manner': Van Slyke, p. 26.

5 Voyages of the Eunuch Admiral

Epigraph: R. Kipling, 1994. *The Collected Poems of Rudyard Kipling*. Wordsworth, Ware, p. 764.

'vessels filled with pearls and precious stones': Needham et al. (1971), p. 530.

'oceanic experience was not part': Yoshihara & Holmes, p. 22.

'Of all the waters under Heaven': Needham et al. (1971), p. 549.

'are made with chambers': ibid., pp. 421–2 (modern translation).

'The boats of the Yellow River': Elvin, p. 141.

'As an engine for carrying men': H. W. Smyth, 1906. *Mast and Sail in Europe and Asia*. J. Murray, London, p. 45.

'There were seven hundred great ships': Watson, p. 138.

'The ships which sail the southern sea': Needham et al. (1971), p. 464.

'deserted the norms of conduct': Keay, p. 372.

'as loud as a huge bell': Levathes, p. 64.

'Bearing vast amounts of gold': Needham & Wang, p. 557.

'the Ming navy probably outclassed': Needham et al. (1971), p. 484.

'have beheld in the ocean': Needham & Wang, p. 558.

'a work of sheer fiction': see http://www.1421exposed.com/html/library_of_congress.html

'a mountain of evidence is accumulating': Needham et al. (1971), p. 545.

'an urbane but systematic tour': ibid., p. 529.

'In Arabia they conversed': ibid., p. 522.

'calm and pacific': ibid., p. 535.

'excellent materials for conducting': Wade, p. 1.

'Zheng He's legacy now serves': Yoshihara & Holmes, p. 110.

'to go to the . . . countries and confer': Keay, p. 384.

'Those who refused submission': Needham & Ronan (1986), p. 142.

'was frightening enough': Dreyer, p. xiv.

'went in succession to the various': ibid., p. 33.

'power projection . . . to force the states': ibid., p. xii.

'intended to create legitimacy': Wade, p. 11.

'I send eunuchs Zheng He': Levathes, p. 169.

'deceitful exaggerations of bizarre things': Needham et al. (1971), p. 525.

'they did not make up for the wasteful': Dreyer, p. 34.

6 Rise and Fall of the Hydraulic State

Epigraph: Elvin, pp. 68–9.

'Without tranquillity': Dodgen, p. 67.

'Now, after more than thirty years': ibid., p. 68.

'He no sooner takes action': ibid.

'They lack tranquility': ibid.

'I rolled the scrolls': ibid.

'By binding their strategic well-being': ibid., pp. 2–3.

'During twenty centuries': Needham et al. (1971), p. 236.

'building embankments on the Yellow River': Chi, p. 73.

'Dykes unmanned': Huang, p. 39.

'guiding the stream in accord': Dodgen, p. 82.

'clothes thin and bellies sunken': ibid., p. 127.

'Year after year, military and river repair': ibid., p. 137.

'Although a gloss of normalcy was restored': ibid., p. 141.

'The British are merely a threat': Lynn, p. 23.

'taking to the land when least expected': Spence, p. 168.

'a tissue of superstition and nonsense': ibid., p. 198.

'peculiar and unintelligible': ibid., p. 207.

'gaudy colours of the dresses': Blakiston, p. 13.

'Things are governed in China': ibid., p. 53.

'Far off, the clouds gathered': Spence, p. 190.

'The devastation is now widespread': Lynn, p. 33.

'Deserted streets': ibid., pp. 33–4.

'the principal emporium': ibid., p. 12.

'many entire villages half-buried': Pietz (2014), p. 73.

7 War on the Waters

Epigraph: Sawyer, p. 284.

'Military tactics are like unto water': Sun Tzu, *The Art of War*, Book VI, 29–32. Transl. L. Giles. http://classics.mit.edu/Tzu/artwar.html

'After crossing a river': ibid., Book IX, 3.

'The main foundations of every state': Niccolò Machiavelli 1961. *The Prince*. Book XII, transl. G. Bull, Penguin, Harmondsworth, p. 77.

'The river is wide, and the tides': *Romance of the Three Kingdoms*, transl. C. H. Brewitt-Taylor, pp. 113–14. https://ebooks.adelaide.edu.au/l/literature/chinese/romance-of-the-three-kingdoms/

'When [the boats] got among the waves': ibid., p. 125.

'The fore parts of the ships': ibid., p. 140.

'When the ships were about a mile': ibid., p. 156.

'These ships have three decks': Needham & Ronan (1986), p. 259.

'Our ships rushed forth': J. Needham, P.-Y. Ho, G.-D. Lu & L. Wang, 1986. *Science and Civilization in China, Vol. V, Part 7: Military Technology; The Gunpowder Epic.* Cambridge University Press, Cambridge, p. 166.

'All ocean-going junks are to be burned': Marks, p. 167.

'Our Celestial Empire possesses all things': H. Scott (ed.), 2015. *The Oxford Handbook of Early Modern European History, 1350–1750: Cultures.* Vol. II, p. 341.

'Two big, lofty, white hulks': Lynn, p. 5.

'The building of dykes': Chi, p. 64.

'greedy and perverse': Sawyer, p. 257.

'I did not originally know': ibid., p. 263.

'Generals confronted by entrenched enemies': ibid.

'Now water is the softest and weakest': ibid., p. 277.

'Using water to assist an attack': ibid., p. 276.

'into fishes, making them live': ibid., p. 278.

'engulf cities, inundate armies': ibid., p. 280.

'lacks the courage': ibid., p. 309.

'When you want to seize the enemy's strength': ibid., p. 317.

'if the colour is black': ibid., p. 342.

8 Mao's Dams

Epigraph: Shapiro (2001), p. vii.

'I have just drunk the waters of Changsha': Lynn, pp. 120–1.

'China will be a green land': Clark, p. 51.

'What we need to learn from Europe': Creel, p. 254.

'If the water power in the Yangtze': Chetham, p. 117.

'How can I face my people': Pietz (2014), p. 78.

'The Hwang Ho and other rivers': ibid., p. 99.

'Our achievements will be nothing less': ibid., p. 134.

'I fear that we have made a mistake': Dai (1997), Ch. 2. https://journal.probeinternational.org/1998/11/30/excerpt-book-river-dragon-has-come-history-dams-china-1/

'chop off the scales': Pietz (2014), p. 217.

'When a sage appears': ibid., p. 227.

'distorted the laws of nature': Shapiro (2001), p. 52.

'Of all China's leaders': ibid, p. 59.

'Accomplishing this pioneering engineering feat': Pietz (2014), p. 149.

'history and culture served': Keay, p. 529.

'Long, long ago, the people of China's': Huang, p. 2.

'In the past, the scourge': Pietz (2014), p. 159.

'Throughout history, the reactionary ruling classes': Huang, p. 12.

'A peasant in his seventies': ibid., p. 73.

'The Three Gorges dam will show the rest': Tvedt & Jakobsson, p. 129.

'any carelessness or negligence': B. Kennedy, 1999, 'China's Three Gorges Dam', CNN. http://edition.cnn.com/SPECIALS/1999/china.50/asian.superpower/three.gorges/

'All the ingredients familiar': Dai (1989), Preface.

'Is this not a vivid expression': Chetham, p. 244.

'chaotic collections of stucco': ibid., p. 2.

'widespread exhaustion of spirit': ibid., p. 12.

'The truth is that the disruption of the dam': Hessler, pp. 108–9.

'I never heard a single resident': ibid., p. 110.

'the common people of China': ibid., p. 106.

9 *The Fluid Art of Expression*

Epigraph: Minford & Lau, p. 799.

'Wind of ruin, lurking rain': Owen, p. 193.

'like the highest mountain': ibid, p. 55.

'read ten thousand volumes': Ward, p. 20.

'What passes. . . . day and night': Allan (1997), p. 11.

'How many dynasties have risen': Lynn, p. 30.

'Go and ask the Yangtze': V. Seth, 2002. 'Poems from Three Chinese Poets'. *India International Centre Quarterly* 29 (2), pp. 81–6.

'Your lonely sail a distant gleam': Watson, p. 144.

'That gleam of sail and mast': ibid.

'banished angel': Liu, p. 79.

'at the river's edge': Lynn, p. 92.

'Do you not see the Waters': Liu, p. 74.

'The water is deep': Hawkes, p. 92.

'a mythological frontier': Van Slyke, p. 138.

'Sometimes intense like myriad pearls': Liu, p. 177.

'Last night the wind raged': Lynn, p. 47.

'*Yin* in its highest form is freezing': 'Yinyang', Internet Encyclopedia of Philosophy. http://www.iep.utm.edu/yinyang/#H5

'During dark or rainy days': Temple, p. 150.

'Human beings are born': Palmer, p. 188.

'Like everything else written': Sze, p. xvii.

'The representation of things necessarily': Bush & Shih, p. 54.

'lively water': Watson, p. 22.

'The quality of the tea': J. Ma, p. 79.

'Wang Wei said': Sze, p. 205.

'causes the water to turn': Bush & Shih, p. 236.

'who paint a waterfall for each fold': ibid., p. 267.

'Clouds are the ornaments of sky': Sze, p. 217.

'Clouds, waterfalls and rivers': Clarke, p. 173.

'Is not water, whether trickling': Sze, p. 203.

'The immensity of the world': Wiseman & Liu, p. 229.

'I know from my own perceptions': Hay, p. 214.

'First, the transformations of water': ibid., p. 271.

'flowing deep down like water': ibid., p. 212.

'Whenever he wanted to paint': 'The Palette of Chinese Painting'; see http://
www.kyfineart.com/index.php?_p=exhibition_info&id=4

'I sit alone in the dark bamboo grove': Liu, p. 72.

'Just as a painter creates his landscape': ibid.

'Scholars who cherish great thoughts': ibid., p. 137.

'Nowadays, those painters who fix titles': Bush & Shih, p. 200.

'A great mountain is dominating': ibid., p. 153.

'The growth in popularity': Murck, p. 124.

'The water has gradually been returned': Hay, p. 78.

'there were vast waves': ibid., pp. 78–9.

'Heaven loves to see people born': ibid., p. 79.

'a world on the verge of dissolution': Clarke, p. 180.

'much water-themed contemporary art': D. Clarke, 2006. 'The watery turn
in contemporary Chinese art'. *Art Journal*, Winter, 55–77, here p. 68.

'there must have been a sense': Clarke, p. 240.

'don't know what their fate will be': Hung, p. 28.

'Although it's still called': Hung, p. 12.

'It felt like these things were being snatched': ibid., p. 17.

'It is not simplistically right or wrong': in *Two Chinas: Chen Qiulin &
Yun-Fei Ji*, catalogue. Worcester Art Museum, Ma., 2008.

'There are two periods': ibid.

10 Water and China's Future

Epigraph: Boulanger, p. 252.

'the abuse of people and abuse of nature': Shapiro (2001), p. xiv.

'The solution for the Sage': Y.-l. Fung, 1952–3. *A History of Chinese Philosophy*,
transl. D. Bodde, Vol. 1. George Allen & Unwin, London, p. 167.

'people who are of ruling quality': Edmonds, p. 24.

'Humanity follows the Earth': see http://www.arcworld.org/faiths.
asp?pageID=11

'At every turn': Shapiro (2001), p. 91.

'working with the Chinese Daoist Association': http://www.arcworld.org/
projects.asp?projectID=266

'Man must use natural science to understand': Economy, p. 48.

'Man's ability to know and change nature': Shapiro (2001), p. 68.

'Everyone is a soldier': Pietz (2014), p. 210.

'Since the dawn of history': Chetham, p. 178.

'it appears that the Lower Yellow River's': A. Borthwick, 'Is the Lower Yellow River sustainable?'. http://www.soue.org.uk/souenews/issue4/yellowriver.html

'The people were deprived': Chi, p. 136.

'To struggle with the heavens': Shapiro (2001), p. 128.

'against revisionism in our minds': ibid.

'There is no Heavenly Emperor or Dragon King': see ibid., p. 68.

'pollution of the environment': Gleick, p. 81.

'Serious water pollution': ibid., p. 97.

'The environment will be the area': Economy, p. 26.

'present a unifying focal point': N. Nankivell, 2005. 'The National Security implications of China's emerging water crisis', Jamestown Foundation, 1 September. http://www.asianresearch.org/articles/2694.html

'Pollution has become the "primer"': Economy, p. 91.

'Inside the system': Larson, 2009.

'most progressive and stringent': Zhang & Cao, p. 433.

'fails to acknowledge citizens' basic': ibid., p. 434.

'fragmented authoritarianism': Lieberthal & Oksenberg. See also K. E. Brosdgaard, 2016. *Chinese Politics as Fragmented Authoritarianism*. Routledge, London.

'I found the chief guarantee': Economy, p. 147.

Bibliography

Allan, S., 1997. *The Way of Water and the Sprouts of Virtue*. SUNY Press, Albany.

Allan, S., 2003. 'The Great One, water, and the Laozi: new light from Guodian', *T'oung Pao* 89, 237–85.

Barnett, J., Rogers, S., Webber, M., Finlayson, B. & Wang, M., 2015. 'Transfer project cannot meet China's water needs', *Nature* 527, 295–7.

Blakiston, T. W., 1862. *Five Months on the Yang-Tsze*. John Murray, London.

Blumenberg, H., 1985. *Work on Myth*, transl. R. M. Wallace. MIT Press, Cambridge, Ma.

Boulanger, N. A., 1764. *The Origin and Progress of Despotism: in the Oriental and Other Empires of Africa, Europe and America*. Amsterdam.

Burke III, E. & Pomeranz, K., 2009. *The Environment and World History*. University of California Press, Berkeley.

Bush, S. & Shih, H.-y., 1985. *Early Chinese Texts on Painting*. Harvard University Press, Cambridge, Ma.

Chan, W.-T. (transl. & ed.), 1963. *A Source Book in Chinese Philosophy*. Princeton University Press, Princeton.

Chang, K. C., 1983. *Art, Myth, and Ritual: The Path to Political Authority in Ancient China*. Harvard University Press, Cambrige, Ma.

Chang, P.-t., 2012. 'The rise of Chinese mercantile power in maritime southeast Asia, *c.*1400–1700', in *Crossroads* 6, http://www.eacrh.net/ojs/index.php/crossroads/article/view/30/Vol6_Chang_html.

Chetham, D., 2002. *Before the Deluge: The Vanishing World of the Yangtze's Three Gorges*. Palgrave Macmillan, New York.

Chi, C.-T., 1936. *Key Economic Areas in Chinese History*. Allen & Unwin, London.

China Education and Research Network, 'Cleaning the Yellow River', 1 January 2001. http://www.edu.cn/special_1506/20060323/t20060323_4751.shtml

Chiu, M. & Zheng, S., 2008. *Art and China's Revolution*. Asia Society, New York & Yale University Press, New Haven.

Clark, C., 1983. *Flood*. Time-Life, Amsterdam.

Clarke, D., 2010. *Water and Art*. Reaktion, London.

Creel, H. G., 1962. *Chinese Thought*. Methuen, London.

Dai, Q. (ed.), 1989. *Yangtze! Yangtze!* Available at https://journal.probeinternational.org/three-gorges-probe/yangtze-yangtze/

Dai, Q. (ed.), 1997. *The River Dragon Has Come!* Available at https://journal.probeinternational.org/the-river-dragon-has-come/

De Villiers, M., 2000. *Water*. Houghton Mifflin, Boston.

Diamond, J., 1998. *Guns, Germs and Steel*. Vintage, London.

Dodgen, R. A., 2001. *Controlling the Dragon: Confucian Engineers and the Yellow River in Late Imperial China*. University of Hawaii Press, Honolulu.

Dreyer, E. L., 2007. *Zheng He: China and the Oceans in the Early Ming Dynasty, 1405–1433*. Pearson Education & Longman, New York.

Dundes, A. (ed.), 1988. *The Flood Myth*. University of California Press, Berkeely.

Economy, E. C., 2010. *The River Runs Black*. Cornell University Press, Ithaca.

Edmonds, R. L., 1994. *Patterns of China's Lost Harmony*. Routledge, London.

Elvin, M., 1973. *The Pattern of the Chinese Past*. Eyre Methuen, London.

Gleick, P. H., 2008. 'China and water'. In P. H. Gleick & M. J. Cohen (eds), *The World's Water 2008–2009*. Island, Washington DC.

Graham, A. C., 1989. *Disputers of the Tao: Philosophical Argument in Ancient China*. Open Court, La Salle, Il.

Greer, C., 1979. *Water Management in the Yellow River Basin of China*. University of Texas Press, Austin.

Hawkes, D., 1987. *A Little Primer of Tu Fu*. Renditions, Hong Kong.

Hay, J., 2001. *Shitao: Painting and Modernity in Early Qing China*. Cambridge University Press, Cambridge.

Hessler, P., 2001. *River Town*. John Murray, London.

Hinton, H. C., 1956. *The Grain Tribute System of China (1845–1911)*. Harvard University Press, Cambridge, Ma.

Huang, W., 1978. *Conquering the Yellow River*. Foreign Languages Press, Beijing.

Hung, W., 2008. *Displacement: The Three Gorges Dam and Contemporary Chinese Art*. Smart Museum of Art, University of Chicago.

Institute for Governance and Sustainable Development, 2010. 'Retreat of Tibetan plateau glaciers caused by global warming threatens water supply and food security'. www.igsd.org/documents/TibetanPlateauGlaciers Note_10August2010.pdf

Jacobs, A., 2013. 'Plans to harness Chinese river's power threaten a region', *New York Times* 4 May. http://www.nytimes.com/2013/05/05/world/asia/plans-to-harness-chinas-nu-river-threaten-a-region.html

Keay, J., 2008. *China: A History*. HarperCollins, London.

Kirk, G. S., 1970. *Myth: Its Meaning and Functions in Ancient and Other Cultures*. Cambridge University Press, London.

C. Larson, 2009. 'On Chinese water project, a struggle over sound science', Environment 360 report http://e360.yale.edu/content/feature.msp?id=2103

Le Comte, Lewis, 1739. *A Compleat History of the Empire of China*. James Hodges, London.

Levathes, L., 1994. *When China Ruled the Seas*. Simon & Schuster, New York.

Lewis, M. E., 2006. *The Flood Myths of Early China*. SUNY Press, Albany.

Li, C.-y., 2012. 'Beneficiary pays: forging reciprocal connections between private profit and public good in hydraulic reform in the Lower Yangzi delta, 1520s-1640s', *T'oung Pao* 98, 385–438.

Li, C.-y., 2015. 'Water management and the legitimization of the Yongle reign, 1403–1424: an approach of political ecology', in *Local Realities and Environmental Changes in the History of East Asia*, ed. T.-j. Liu, pp. 51–87. Routledge, London.

Li, C.-y., 2010. 'Contending strategies, collaboration among local specialists and officials, and hydrological reform in the late-fifteenth-century Lower Yangzi delta', *East Asian Science, Technology and Society* 4, 229–53.

Lieberthal, K. & Oksenberg, M. (eds), 1988. *Policy Making in China: Leaders, Structures, and Processes*. Princeton University Press, Princeton.

Liu, W.-C., 1966. *An Introduction to Chinese Literature*. Indiana University Press, Bloomington.

Lorge, P., 2005. *War, Politics and Society in Early Modern China 900–1795*. Routledge, London.

Lynn, M. (ed.), 1997. *Yangtze River: The Wildest, Wickedest River on Earth*. Oxford University Press, Oxford.

Ma, J., 2004. *China's Water Crisis*, transl. N. Y. Liu & L. R. Sullivan. Eastbridge, Norwalk, Ct.

Marks, R. B., 2011. *China: Its Environment and History*. Rowman & Littlefield, Lanham, Md.

Marks, R. B., 1998. *Tigers, Rice, Silk, and Silt: Environment and Economy in Late Imperial South China*. Cambridge University Press, Cambridge.

Martin, B. G., 1996. *The Shanghai Green Gang: Politics and Organized Crime, 1919–1937*. University of California Press, Berkeley.

Mathews, J. A. & Tan, H., 2014. 'Manufacture renewables to build national security', *Nature* 513, 166.

Mertha, A. C., 2010. *China's Water Warriors: Citizen Action and Policy Change*. Cornell University Press, Ithaca, NY.

Minford, J. & Lau, J. S. M. (eds), 2000. *Chinese Classical Literature: An Anthology of Translations, Vol. I: From Antiquity to the Tang Dynasty*. Columbia University Press, New York.

Mithen, S., 2012. *Thirst: Water and Power in the Ancient World*. Weidenfeld & Nicolson, London.

Murck, A., 2000. *Poetry and Painting in Song China: The Subtle Art of Dissent.* Harvard University Press, Cambridge, Ma.

Needham, J. & Ronan, C. A., 1978. *The Shorter Science and Civilization in China*, Vol. 1. Cambridge University Press, Cambridge.

Needham, J. & Ronan, C. A., 1986. *The Shorter Science and Civilisation in China*, Vol. 3. Cambridge University Press, Cambridge.

Needham, J. & Wang, L., 1959. *Science and Civilization in China, Vol. III: Mathematics and the Sciences of the Heavens and the Earth.* Cambridge University Press, Cambridge.

Needham, J., Wang, L. & Lu, G.-D., 1971. *Science and Civilization in China, Vol. IV, Part 3: Physics and Physical Technology: Civil Engineering and Nautics.* Cambridge University Press, Cambridge.

Oldstone-Moore, J., 2003. *Understanding Confucianism.* Duncan Baird, London.

Owen, S., 1981. *The Great Age of Chinese Poetry: The High T'ang.* Yale University Press, New Haven.

Padovani, F., 2006. 'The Chinese way of harnessing rivers: the Yangtze River'. In T. Tvedt & E. Jakobsson (eds), *A History of Water, Vol. 1: Water Control and River Biographies,* pp. 120–43. I. B. Tauris, London.

Padovani, F., 2006. 'Displacement from the Three Gorges region', *China Perspectives,* July–August.

Palmer, M. (transl.), 2006. *The Book of Chuang Tzu.* Penguin, London.

Palmer, M. A., Liu, J., Matthews, J. H., Mumba, M. & D-Odorico, P., 2015. 'Water security: gray or green?', *Science* 349, 584–5.

Perdue, P. C., 1987. *Exhausting the Earth: State and Peasant in Hunan, 1500–1850.* Harvard University Press, Cambridge.

Pietz, D., 2006. 'Controlling the waters in twentieth-century China: The Nationalist state and the Huai River'. In T. Tvedt & E. Jakobsson (eds), *A History of Water, Vol. 1: Water Control and River Biographies,* pp. 92–119. I. B. Tauris, London.

Pietz, D., 2014. *The Yellow River: The Problem of Water in Modern China.* Harvard University Press, Cambridge, Ma.

Priscoli, J. D. & Wolf, A. T., 2009. *Managing and Transforming Water Conflicts.* Cambridge University Press, New York.

Qiu, J., 2014. 'Chinese data hint at trigger for fatal quake', *Nature* 513, 154.

Rawski, T. G. & Li, L. M., 1992. *Chinese History in Economic Perspective.* University of California Press, Berkeley.

Ronan, C. A., 1986. *The Shorter Science and Civilisation in China*, Vol. 3. Cambridge University Press, Cambridge.

Ropp, P. S. (ed.), 1990. *Heritage of China.* University of California Press, Berkeley.

Sage, S. F., 1992. *Ancient Sichuan and the Unification of China.* SUNY Press, Albany.

Sawyer, R. D., 2004. *Fire and Water: The Art of Incendiary and Aquatic Warfare in China*. Westfield, Boulder, Co.

Schafer, E. H., 1973. *The Divine Woman: Dragon Ladies and Rain Maidens in T'ang Literature*. University of California Press, Berkeley.

Schoppa, R. K., 1989. *Xiang Lake: Nine Centuries of Chinese Life*. Yale University Press, New Haven.

Shang, Y., Shang, H., Liang, J., Shen, H. & Liu, G., 2013. 'Comprehensive study on degradation and management of Baiyangdian Lake in North China', *Journal of Environmental Science and Engineering B* 2, 337–42.

Shapiro, J., 2001. *Mao's War Against Nature*. Cambridge University Press, Cambridge.

Shapiro, J., 2012. *China's Environmental Challenges*. Polity, Cambridge.

Shaughnessey, E. L., 2005. *Ancient China: Life, Myth and Art*. Duncan Baird, London.

Spence, J., 1996. *God's Chinese Son: The Taiping Heavenly Kingdom of Hong Xiuquan*. W. W. Norton, New York.

Spence, J. D., 1999. *The Search for Modern China*. W. W. Norton, New York.

Strang, V., 2004. *The Meaning of Water*. Berg, Oxford.

Sze, M.-m. (ed. & transl.), 1978. *The Mustard Seed Garden Manual of Painting*. Princeton University Press, Princeton.

Talhelm, T., Zhang, X., Oishi, S., Shimin, C., Duan, D., Lan, X. & Kitayama, S., 2014. 'Large-scale psychological differences within China explained by rice versus wheat agriculture', *Science* 344, 603–608.

Temple, R., 1998. *The Genius of China*. Prion, London.

Van Slyke, L. P., 1988. *Yangtze: Nature, History and the River*. Addison Wesley, Reading, Ma.

Viollet, P.-L., 2007. *Water Engineering in Ancient Civilizations: 5,000 Years of History*. CRC Press, Boca Raton, Fl.

Wade, G., 2004. 'The Zheng He voyages: a reassessment', Asia Research Institute Working Paper No. 31. http://www.ari.nus.edu.sg/wps/wps04_031.pdf

Ward, J., 2001. *Xu Xiake (1587–1641): The Art of Travel Writing*. Curzon, Richmond.

Watson, P. (transl.), 2007. *Grand Canal, Great River: The Travel Diary of a Twelfth-Century Chinese Poet*. Francis Lincoln, London.

Watts, A., 1976. *Tao: The Watercourse Way*. Jonathan Cape, London.

Wilkinson, P., 2005. *Yangtze*. BBC Books, London.

Winchester, S., 1998. *The River at the Centre of the World*. Penguin, London.

Wiseman, M. B. & Liu, Y., 2011. *Subversive Strategies in Contemporary Chinese Art*. Brill, Leiden.

Wittfogel, K. A., 1957. *Oriental Despotism: A Comparative Study of Total Power*. Yale University Press, New Haven.

Wu, H., 2008. *Displacement: The Three Gorges Dam and Contemporary Chinese Art*. Smart Museum of Art, University of Chicago.

Wu, K. C., 1982. *The Chinese Heritage*. Crown, New York.

Xiong, V. C., 2006. *Emperor Yang of the Sui Dynasty: His Life, Times, and Legacy*. SUNY Press, Albany.

Xu, J., 2005. 'The water fluxes of the Yellow River to the sea in the past 50 years, in response to climate change and human activities', *Environmental Management* 35, 620–31.

Yardley, J., 2007. 'Chinese dam projects criticized for their human costs', *New York Times* 19 November. http://www.nytimes.com/2007/11/19/world/asia/19dam.html

Yoshihara, T. & Holmes, J. R. (eds), 2008. *Asia Looks Seaward: Power and Maritime Strategy*. Prager Security International, Westport, Ct.

Zheng Xiao Yun, 2010. 'Shaping beliefs, identities and institutions: the role of water myths among ethnic groups in Yunnan, China'. In T. Tvedt & T. Oestigaard (eds), *A History of Water. Series II, Vol. 1: Ideas of Water from Ancient Societies to the Modern World*, pp. 405–423. I. B. Tauris, London.

Zhang, B. & Cao, C., 2015. 'Four gaps in China's new environmental law', *Nature* 517, 433–4.

Zhang, Q., Xu, C.-Y., Yang, T. & Hao, Z.-C., 2010. 'The historical evolution and anthropogenic influences on the Yellow River from ancient times to modern times.' In T. Tvedt & R. Coopey (eds), *A History of Water. Series II, Vol. 2: Rivers and Society: From Early Civilizations to Modern Times*, pp. 144–64. I. B. Tauris, London.

Index